>SPSS 16.0 Brief Guide

For more information about SPSS® software products, please visit our Web site at *http://www.spss.com* or contact

SPSS Inc.
233 South Wacker Drive, 11th Floor
Chicago, IL 60606-6412
Tel: (312) 651-3000
Fax: (312) 651-3668

SPSS is a registered trademark and the other product names are the trademarks of SPSS Inc. for its proprietary computer software. No material describing such software may be produced or distributed without the written permission of the owners of the trademark and license rights in the software and the copyrights in the published materials.

The SOFTWARE and documentation are provided with RESTRICTED RIGHTS. Use, duplication, or disclosure by the Government is subject to restrictions as set forth in subdivision (c) (1) (ii) of The Rights in Technical Data and Computer Software clause at 52.227-7013. Contractor/manufacturer is SPSS Inc., 233 South Wacker Drive, 11th Floor, Chicago, IL 60606-6412.
Patent No. 7,023,453

General notice: Other product names mentioned herein are used for identification purposes only and may be trademarks of their respective companies.

Windows is a registered trademark of Microsoft Corporation.

Apple, Mac, and the Mac logo are trademarks of Apple Computer, Inc., registered in the U.S. and other countries.

This product uses WinWrap Basic, Copyright 1993-2007, Polar Engineering and Consulting, *http://www.winwrap.com.*

SPSS 16.0 Brief Guide
Copyright © 2007 by SPSS Inc.
All rights reserved.
Printed in the United States of America.

No part of this publication may be reproduced, stored in a retrieval system, or transmitted, in any form or by any means, electronic, mechanical, photocopying, recording, or otherwise, without the prior written permission of the publisher.

ISBN-13: 978-0-13-603601-2
ISBN-10: 0-13-603601-5

1 2 3 4 5 6 7 8 9 0   10 09 08 07

Preface

The *SPSS 16.0 Brief Guide* provides a set of tutorials designed to acquaint you with the various components of SPSS. This guide is intended for use with all operating system versions of the software, including: Windows, Macintosh, and Linux. You can work through the tutorials in sequence or turn to the topics for which you need additional information. You can use this guide as a supplement to the online tutorial that is included with the SPSS Base 16.0 system or ignore the online tutorial and start with the tutorials found here.

SPSS 16.0

SPSS 16.0 is a comprehensive system for analyzing data. SPSS can take data from almost any type of file and use them to generate tabulated reports, charts, and plots of distributions and trends, descriptive statistics, and complex statistical analyses.

SPSS makes statistical analysis more accessible for the beginner and more convenient for the experienced user. Simple menus and dialog box selections make it possible to perform complex analyses without typing a single line of command syntax. The Data Editor offers a simple and efficient spreadsheet-like facility for entering data and browsing the working data file.

Internet Resources

The SPSS Web site (*http://www.spss.com*) offers answers to frequently asked questions about installing and running SPSS software and provides access to data files and other useful information.

In addition, the SPSS USENET discussion group (not sponsored by SPSS) is open to anyone interested in SPSS products. The USENET address is *comp.soft-sys.stat.spss*. It deals with computer, statistical, and other operational issues related to SPSS software.

You can also subscribe to an e-mail message list that is gatewayed to the USENET group. To subscribe, send an e-mail message to *listserv@listserv.uga.edu*. The text of the e-mail message should be: subscribe SPSSX-L firstname lastname. You can then post messages to the list by sending an e-mail message to *listserv@listserv.uga.edu*.

Additional Publications

For additional information about the features and operations of SPSS Base 16.0, you can consult the *SPSS Base 16.0 User's Guide*, which includes information on standard graphics. Examples using the statistical procedures found in SPSS Base 16.0 are provided in the Help system, installed with the software. Algorithms used in the statistical procedures are available on the product CD-ROM.

In addition, beneath the menus and dialog boxes, SPSS uses a command language. Some extended features of the system can be accessed only via command syntax. (Those features are not available in the Student Version.) Complete command syntax is documented in the *SPSS 16.0 Command Syntax Reference*, available in PDF form from the Help menu.

Individuals worldwide can order additional product manuals directly from SPSS Inc. For telephone orders in the United States and Canada, call SPSS Inc. at 800-543-2185. For telephone orders outside of North America, contact your local office, listed on the SPSS Web site at *http://www.spss.com/worldwide*.

The *SPSS Statistical Procedures Companion*, by Marija Norušis, has been published by Prentice Hall. It contains overviews of the procedures in SPSS Base, plus Logistic Regression and General Linear Models. The *SPSS Advanced Statistical Procedures Companion* has also been published by Prentice Hall. It includes overviews of the procedures in the Advanced and Regression modules.

SPSS Options

The following options are available as add-on enhancements to the full (not Student Version) SPSS Base system:

SPSS Regression Models™ provides techniques for analyzing data that do not fit traditional linear statistical models. It includes procedures for probit analysis, logistic regression, weight estimation, two-stage least-squares regression, and general nonlinear regression.

SPSS Advanced Models™ focuses on techniques often used in sophisticated experimental and biomedical research. It includes procedures for general linear models (GLM), linear mixed models, variance components analysis, loglinear analysis, ordinal regression, actuarial life tables, Kaplan-Meier survival analysis, and basic and extended Cox regression.

SPSS Tables™ creates a variety of presentation-quality tabular reports, including complex stub-and-banner tables and displays of multiple response data.

SPSS Trends™ performs comprehensive forecasting and time series analyses with multiple curve-fitting models, smoothing models, and methods for estimating autoregressive functions.

SPSS Categories® performs optimal scaling procedures, including correspondence analysis.

SPSS Conjoint™ provides a realistic way to measure how individual product attributes affect consumer and citizen preferences. With SPSS Conjoint, you can easily measure the trade-off effect of each product attribute in the context of a set of product attributes—as consumers do when making purchasing decisions.

SPSS Exact Tests™ calculates exact p values for statistical tests when small or very unevenly distributed samples could make the usual tests inaccurate. This option is available only on Windows operating systems.

SPSS Missing Value Analysis™ describes patterns of missing data, estimates means and other statistics, and imputes values for missing observations.

SPSS Complex Samples™ allows survey, market, health, and public opinion researchers, as well as social scientists who use sample survey methodology, to incorporate their complex sample designs into data analysis.

SPSS Classification Tree™ creates a tree-based classification model. It classifies cases into groups or predicts values of a dependent (target) variable based on values of independent (predictor) variables. The procedure provides validation tools for exploratory and confirmatory classification analysis.

SPSS Data Preparation™ provides a quick visual snapshot of your data. It provides the ability to apply validation rules that identify invalid data values. You can create rules that flag out-of-range values, missing values, or blank values. You can also save variables that record individual rule violations and the total number of rule violations per case. A limited set of predefined rules that you can copy or modify is provided.

SPSS Neural Networks™ can be used to make business decisions by forecasting demand for a product as a function of price and other variables, or by categorizing customers based on buying habits and demographic characteristics. Neural networks are non-linear data modeling tools. They can be used to model complex relationships between inputs and outputs or to find patterns in data.

Amos™ (**a**nalysis of **mo**ment **s**tructures) uses structural equation modeling to confirm and explain conceptual models that involve attitudes, perceptions, and other factors that drive behavior.

The SPSS family of products also includes applications for data entry, text analysis, classification, neural networks, and predictive enterprise services.

Training Seminars

SPSS Inc. provides both public and onsite training seminars for SPSS. All seminars feature hands-on workshops. SPSS seminars will be offered in major U.S. and European cities on a regular basis. For more information on these seminars, contact your local office, listed on the SPSS Web site at *http://www.spss.com/worldwide*.

Technical Support

The services of SPSS Technical Support are available to maintenance customers of SPSS. (Student Version customers should read the special section on technical support for the Student Version. For more information, see "Technical Support for Students" on p. viii.) Customers may contact Technical Support for assistance in using SPSS products or for installation help for one of the supported hardware environments. To reach Technical Support, see the SPSS Web site at *http://www.spss.com*, or contact your local office, listed on the SPSS Web site at *http://www.spss.com/worldwide*. Be prepared to identify yourself, your organization, and the serial number of your system.

Tell Us Your Thoughts

Your comments are important. Please let us know about your experiences with SPSS products. We especially like to hear about new and interesting applications using SPSS. Please send e-mail to *suggest@spss.com*, or write to SPSS Inc., Attn: Director of Product Planning, 233 South Wacker Drive, 11th Floor, Chicago IL 60606-6412.

SPSS 16.0 for Windows Student Version

The SPSS 16.0 for Windows Student Version is a limited but still powerful version of the SPSS Base 16.0 system.

Capability

The Student Version contains all of the important data analysis tools contained in the full SPSS Base system, including:

- Spreadsheet-like Data Editor for entering, modifying, and viewing data files.
- Statistical procedures, including t tests, analysis of variance, and crosstabulations.
- Interactive graphics that allow you to change or add chart elements and variables dynamically; the changes appear as soon as they are specified.
- Standard high-resolution graphics for an extensive array of analytical and presentation charts and tables.

Limitations

Created for classroom instruction, the Student Version is limited to use by students and instructors for educational purposes only. The Student Version does not contain all of the functions of the SPSS Base 16.0 system. The following limitations apply to the SPSS 16.0 for Windows Student Version:

- Data files cannot contain more than 50 variables.
- Data files cannot contain more than 1,500 cases. SPSS add-on modules (such as Regression Models or Advanced Models) cannot be used with the Student Version.
- SPSS command syntax is not available to the user. This means that it is not possible to repeat an analysis by saving a series of commands in a syntax or "job" file, as can be done in the full version of SPSS .
- Scripting and automation are not available to the user. This means that you cannot create scripts that automate tasks that you repeat often, as can be done in the full version of SPSS .

Technical Support for Students

Students should obtain technical support from their instructors or from local support staff identified by their instructors. Technical support from SPSS for the SPSS 16.0 Student Version is provided *only to instructors using the system for classroom instruction*.

Before seeking assistance from your instructor, please write down the information described below. Without this information, your instructor may be unable to assist you:

- The type of computer you are using, as well as the amount of RAM and free disk space you have.
- The operating system of your computer.
- A clear description of what happened and what you were doing when the problem occurred. If possible, please try to reproduce the problem with one of the sample data files provided with the program.
- The exact wording of any error or warning messages that appeared on your screen.
- How you tried to solve the problem on your own.

Technical Support for Instructors

Instructors using the Student Version for classroom instruction may contact SPSS Technical Support for assistance. In the United States and Canada, call SPSS Technical Support at (312) 651-3410, or send an e-mail to *support@spss.com*. Please include your name, title, and academic institution.

Instructors outside of the United States and Canada should contact your local SPSS office, listed on the SPSS Web site at *http://www.spss.com/worldwide*.

Contents

4 Working with Multiple Data Sources   59

5 Examining Summary Statistics for Individual Variables   65

6 Creating and Editing Charts   75

7   *Working with Output*   *105*

11 Additional Statistical Procedures 185

Appendix

Introduction

This guide provides a set of tutorials designed to enable you to perform useful analyses on your data. You can work through the tutorials in sequence or turn to the topics for which you need additional information.

This chapter will introduce you to the basic features and demonstrate a typical session. We will retrieve a previously defined SPSS-format data file and then produce a simple statistical summary and a chart.

More detailed instruction about many of the topics touched upon in this chapter will follow in later chapters. Here, we hope to give you a basic framework for understanding later tutorials.

Sample Files

Most of the examples that are presented here use the data file *demo.sav*. This data file is a fictitious survey of several thousand people, containing basic demographic and consumer information.

The sample files installed with the product can be found in the *Samples* subdirectory of the installation directory.

Opening a Data File

To open a data file:

▶ From the menus choose:
File
  Open
    Data...

Alternatively, you can use the Open File button on the toolbar.

Figure 1-1
Open File toolbar button

A dialog box for opening files is displayed.

By default, SPSS-format data files (*.sav* extension) are displayed.

This example uses the file *demo.sav*.

Figure 1-2
demo.sav file in Data Editor

	age	marital	address	income	inccat	car
1	55	Marital status	12	72.00	3.00	36.
2	56	0	29	153.00	4.00	76.
3	28	1	9	28.00	2.00	13.
4	24	1	4	26.00	2.00	12.
5	25	0	2	23.00	1.00	11.
6	45	1	9	76.00	4.00	37.
7	42	0	19	40.00	2.00	19.
8	35	0	15	57.00	3.00	28.
9	46	0	26	24.00	1.00	12.
10	34	1	0	89.00	4.00	46.
11	55	1	17	72.00	3.00	35

demo.sav - Data Editor

File Edit View Data Transform Analyze Graphs Utilities Add-ons Window Help

20 : age   40

Data View / Variable View

The data file is displayed in the Data Editor. In the Data Editor, if you put the mouse cursor on a variable name (the column headings), a more descriptive variable label is displayed (if a label has been defined for that variable).

By default, the actual data values are displayed. To display labels:

▶ From the menus choose:
View
  Value Labels

Alternatively, you can use the Value Labels button on the toolbar.

Figure 1-3
Value Labels button

Descriptive value labels are now displayed to make it easier to interpret the responses.

Figure 1-4
Value labels displayed in the Data Editor

	age	marital	address	income	inccat	car
1	55	Married	12	72.00	$50 - $74	36.
2	56	Unmarried	29	153.00	$75+	76.
3	28	Married	9	28.00	$25 - $49	13.
4	24	Married	4	26.00	$25 - $49	12.
5	25	Unmarried	2	23.00	Under $25	11.
6	45	Married	9	76.00	$75+	37.
7	42	Unmarried	19	40.00	$25 - $49	19.
8	35	Unmarried	15	57.00	$50 - $74	28.
9	46	Unmarried	26	24.00	Under $25	12.
10	34	Married	0	89.00	$75+	46.
11	55	Married	17	72.00	$50 - $74	35.

demo.sav - Data Editor

File Edit View Data Transform Analyze Graphs Utilities Add-ons Window Help

20 : age   40

Data View / Variable View /

Running an Analysis

The Analyze menu contains a list of general reporting and statistical analysis categories.

We will start by creating a simple frequency table (table of counts).

▶ From the menus choose:

Analyze
  Descriptive Statistics
    Frequencies...

The Frequencies dialog box is displayed.

Figure 1-5
Frequencies dialog box

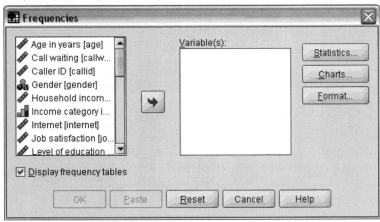

An icon next to each variable provides information about data type and level of measurement.

Measurement Level	Data Type			
	Numeric	String	Date	Time
Scale		n/a		
Ordinal				
Nominal				

▶ Click the variable *Income category in thousands [inccat]*.

Figure 1-6
Variable labels and names in the Frequencies dialog box

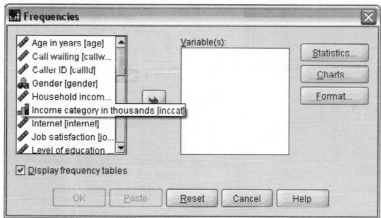

If the variable label and/or name appears truncated in the list, the complete label/name is displayed when the cursor is positioned over it. The variable name *inccat* is displayed in square brackets after the descriptive variable label. *Income category in thousands* is the variable label. If there were no variable label, only the variable name would appear in the list box.

You can resize dialog boxes just like windows, by clicking and dragging the outside borders or corners. For example, if you make the dialog box wider, the variable lists will also be wider.

Figure 1-7
Resized dialog box

In the dialog box, you choose the variables that you want to analyze from the source list on the left and drag and drop them into the Variable(s) list on the right. The OK button, which runs the analysis, is disabled until at least one variable is placed in the Variable(s) list.

You can obtain additional information by right-clicking any variable name in the list.

▶ Right-click *Income category in thousands [inccat]* and choose Variable Information.

▶ Click the down arrow on the Value labels drop-down list.

Figure 1-8
Defined labels for income variable

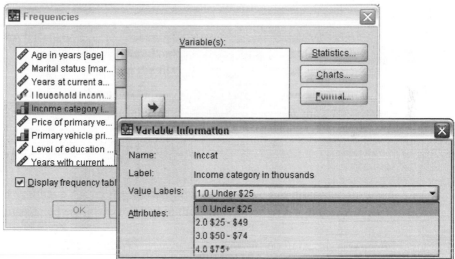

All of the defined value labels for the variable are displayed.

▶ Click *Gender [gender]* in the source variable list and drag the variable into the target Variable(s) list.

▶ Click *Income category in thousands [inccat]* in the source list and drag it to the target list.

Figure 1-9
Variables selected for analysis

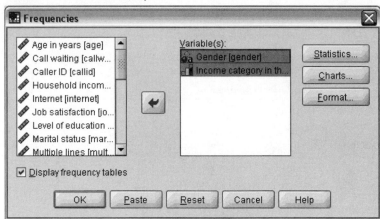

▶ Click OK to run the procedure.

Viewing Results

Figure 1-10
Viewer window

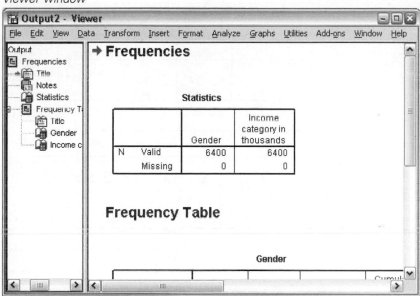

Results are displayed in the Viewer window.

You can quickly go to any item in the Viewer by selecting it in the outline pane.

▶ Click Income category in thousands [inccat].

Figure 1-11
Frequency table of income categories

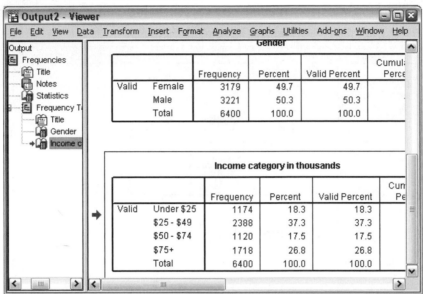

The frequency table for income categories is displayed. This frequency table shows the number and percentage of people in each income category.

Creating Charts

Although some statistical procedures can create charts, you can also use the Graphs menu to create charts.

For example, you can create a chart that shows the relationship between wireless telephone service and PDA (personal digital assistant) ownership.

▶ From the menus choose:
Graphs
   Chart Builder...

▶ Click the Gallery tab (if it is not selected).

▶ Click Bar (if it is not selected).

▶ Drag the Clustered Bar icon onto the canvas, which is the large area above the Gallery.

Figure 1-12
Chart Builder dialog box

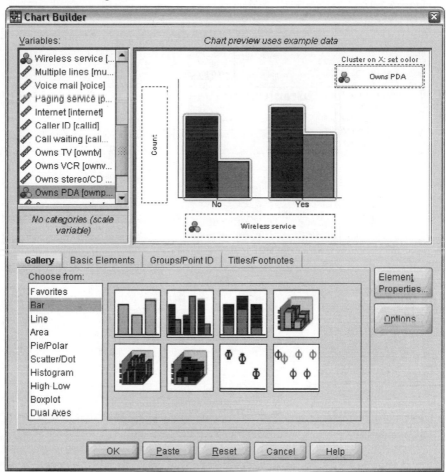

▶ Scroll down the Variables list, right-click *Wireless service [wireless]*, and then choose Nominal as its measurement level.

▶ Drag the *Wireless service [wireless]* variable to the *x* axis.

▶ Right-click *Owns PDA [ownpda]* and choose Nominal as its measurement level.

▶ Drag the *Owns PDA [ownpda]* variable to the cluster drop zone in the upper right corner of the canvas.

▶ Click OK to create the chart.

Figure 1-13
Bar chart displayed in Viewer window

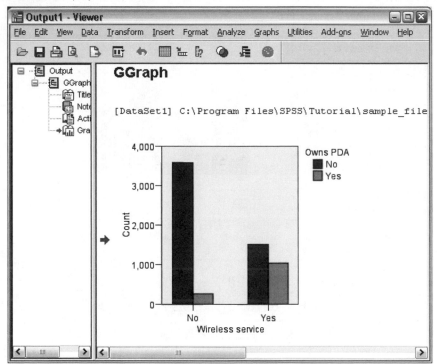

The bar chart is displayed in the Viewer. The chart shows that people with wireless phone service are far more likely to have PDAs than people without wireless service.

You can edit charts and tables by double-clicking them in the contents pane of the Viewer window, and you can copy and paste your results into other applications. Those topics will be covered later.

Chapter
2

Reading Data

Data can be entered directly, or it can be imported from a number of different sources. The processes for reading data stored in SPSS-format data files; spreadsheet applications, such as Microsoft Excel; database applications, such as Microsoft Access; and text files are all discussed in this chapter.

Basic Structure of an SPSS-format Data File

Figure 2-1
Data Editor

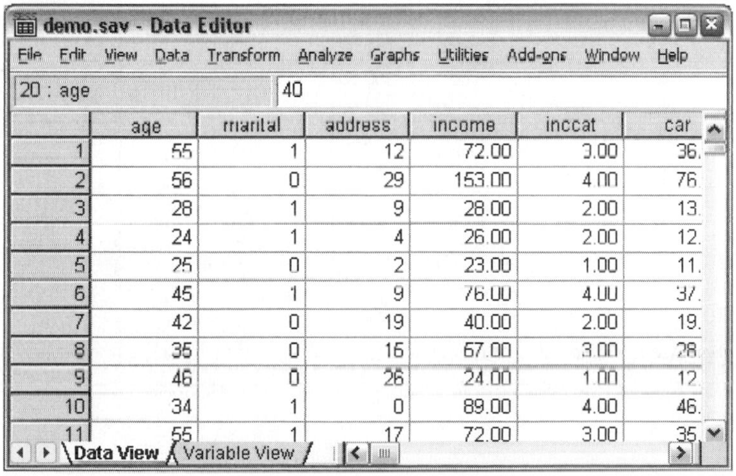
demo.sav - Data Editor
File Edit View Data Transform Analyze Graphs Utilities Add-ons Window Help
20 : age
40
age marital address income inccat car
1 55 1 12 72.00 3.00 36.
2 56 0 29 153.00 4.00 76.
3 28 1 9 28.00 2.00 13.
4 24 1 4 26.00 2.00 12.
5 25 0 2 23.00 1.00 11.
6 45 1 9 76.00 4.00 37.
7 42 0 19 40.00 2.00 19.
8 35 0 16 67.00 3.00 28.
9 46 0 26 24.00 1.00 12.
10 34 1 0 89.00 4.00 46.
11 55 1 17 72.00 3.00 35.
Data View
Variable View

SPSS-format data files are organized by cases (rows) and variables (columns). In this data file, cases represent individual respondents to a survey. Variables represent responses to each question asked in the survey.

Reading an SPSS-format Data File

SPSS-format data files, which have a *.sav* file extension, contain your saved data. To open demo.sav, an example file installed with the product:

▶ From the menus choose:
File
  Open
    Data...

▶ Browse to and open *demo.sav*. For more information, see "Sample Files" in Appendix A on p. 201.

The data are now displayed in the Data Editor.

Figure 2-2
Opened data file

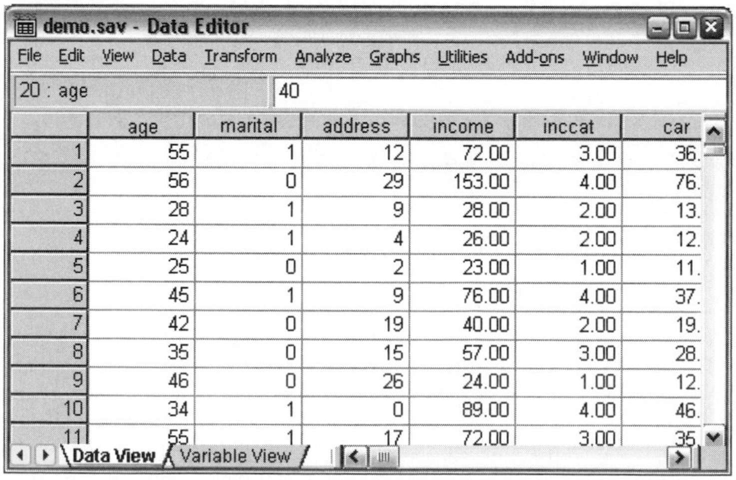

Reading Data from Spreadsheets

Rather than typing all of your data directly into the Data Editor, you can read data from applications such as Microsoft Excel. You can also read column headings as variable names.

▶ From the menus choose:

File
  Open
    Data...

▶ Select Excel (*.xls) as the file type you want to view.

▶ Open *demo.xls*. For more information, see "Sample Files" in Appendix A on p. 201.

The Opening Excel Data Source dialog box is displayed, allowing you to specify whether variable names are to be included in the spreadsheet, as well as the cells that you want to import. In Excel 95 or later, you can also specify which worksheets you want to import.

Figure 2-3
Opening Excel Data Source dialog box

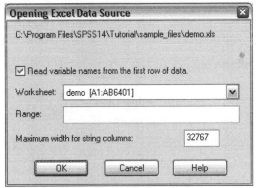

▶ Make sure that Read variable names from the first row of data is selected. This option reads column headings as variable names.

If the column headings do not conform to the SPSS variable-naming rules, they are converted into valid variable names and the original column headings are saved as variable labels. If you want to import only a portion of the spreadsheet, specify the range of cells to be imported in the Range text box.

▶ Click OK to read the Excel file.

The data now appear in the Data Editor, with the column headings used as variable names. Since variable names can't contain spaces, the spaces from the original column headings have been removed. For example, *Marital status* in the Excel file becomes the variable *Maritalstatus*. The original column heading is retained as a variable label.

Figure 2-4
Imported Excel data

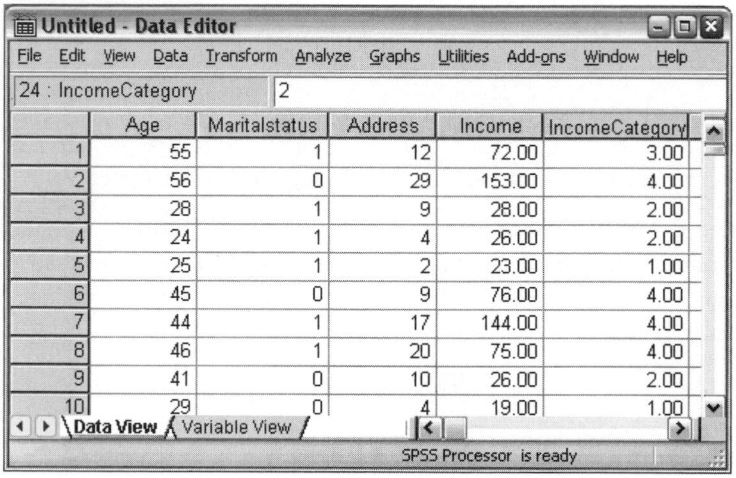

	Age	Maritalstatus	Address	Income	IncomeCategory
1	55	1	12	72.00	3.00
2	56	0	29	153.00	4.00
3	28	1	9	28.00	2.00
4	24	1	4	26.00	2.00
5	25	1	2	23.00	1.00
6	45	0	9	76.00	4.00
7	44	1	17	144.00	4.00
8	46	1	20	75.00	4.00
9	41	0	10	26.00	2.00
10	29	0	4	19.00	1.00

Untitled - Data Editor. File Edit View Data Transform Analyze Graphs Utilities Add-ons Window Help. 24 : IncomeCategory 2. Data View / Variable View. SPSS Processor is ready.

Reading Data from a Database

Data from database sources are easily imported using the Database Wizard. Any database that uses ODBC (Open Database Connectivity) drivers can be read directly after the drivers are installed. ODBC drivers for many database formats are supplied on the installation CD. Additional drivers can be obtained from third-party vendors. One of the most common database applications, Microsoft Access, is discussed in this example.

Note: This example is specific to Microsoft Windows and requires an ODBC driver for Access. The steps are similar on other platforms but may require a third-party ODBC driver for Access.

▶ From the menus choose:
File
  Open Database
    New Query...

Figure 2-5
Database Wizard Welcome dialog box

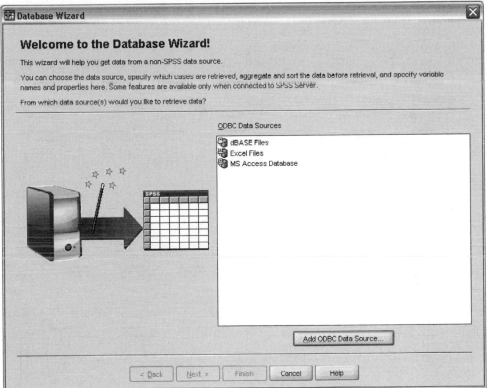

▶ Select **MS Access Database** from the list of data sources and click Next.

Note: Depending on your installation, you may also see a list of OLEDB data sources on the left side of the wizard (Windows operating systems only), but this example uses the list of ODBC data sources displayed on the right side.

Figure 2-6
ODBC Driver Login dialog box

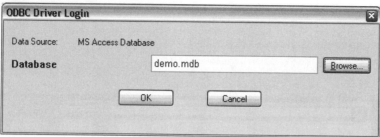

▶ Click Browse to navigate to the Access database file that you want to open.

▶ Open *demo.mdb*. For more information, see "Sample Files" in Appendix A on p. 201.

▶ Click OK in the login dialog box.

In the next step, you can specify the tables and variables that you want to import.

Figure 2-7
Select Data step

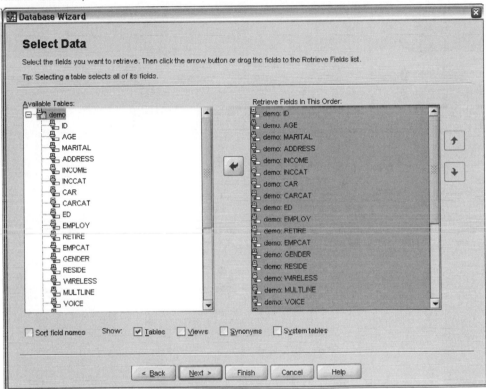

▶ Drag the entire demo table to the Retrieve Fields In This Order list.

▶ Click Next.

In the next step, you select which records (cases) to import.

Figure 2-8
Limit Retrieved Cases step

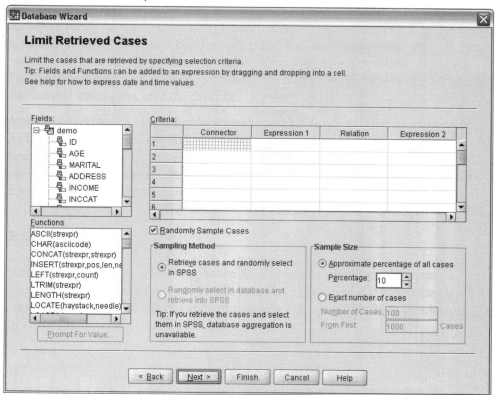

If you do not want to import all cases, you can import a subset of cases (for example, males older than 30), or you can import a random sample of cases from the data source. For large data sources, you may want to limit the number of cases to a small, representative sample to reduce the processing time.

▶ Click Next to continue.

Field names are used to create variable names. If necessary, the names are converted to valid variable names. The original field names are preserved as variable labels. You can also change the variable names before importing the database.

Figure 2-9
Define Variables step

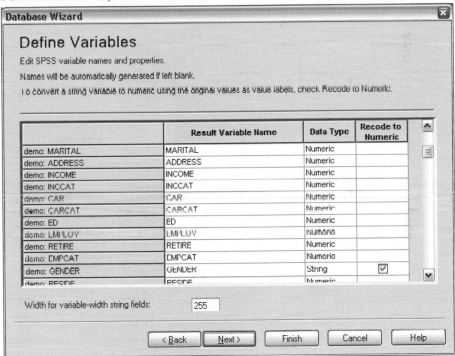

▶ Click the Recode to Numeric cell in the Gender field. This option converts string variables to integer variables and retains the original value as the value label for the new variable.

▶ Click Next to continue.

The SQL statement created from your selections in the Database Wizard appears in the Results step. This statement can be executed now or saved to a file for later use.

Figure 2-10
Results step

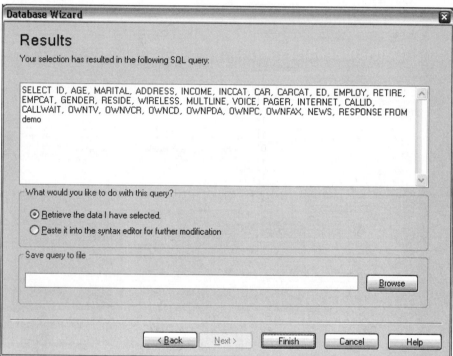

▶ Click Finish to import the data.

All of the data in the Access database that you selected to import are now available in the Data Editor.

Figure 2-11
Data imported from an Access database

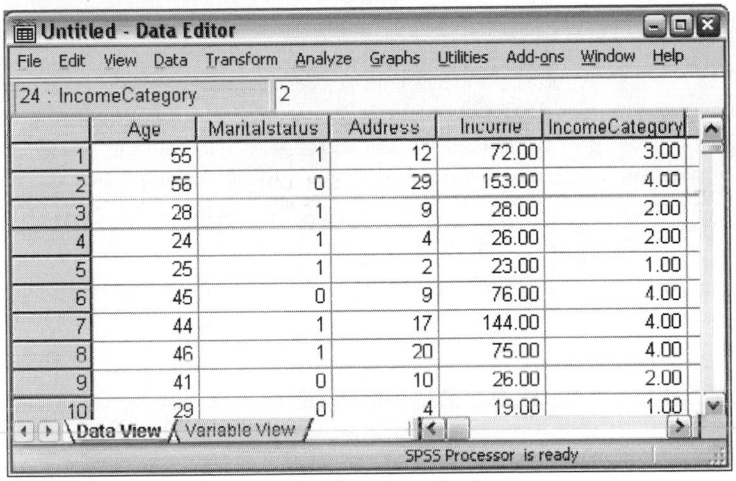

Reading Data from a Text File

Text files are another common source of data. Many spreadsheet programs and databases can save their contents in one of many text file formats. Comma- or tab-delimited files refer to rows of data that use commas or tabs to indicate each variable. In this example, the data are tab delimited.

▶ From the menus choose:
File
  Read Text Data...

▶ Select Text (ʻ.txt) as the file type you want to view.

▶ Open *demo.txt*. For more information, see "Sample Files" in Appendix A on p. 201.

The Text Import Wizard guides you through the process of defining how the specified text file should be interpreted.

Figure 2-12
Text Import Wizard: Step 1 of 6

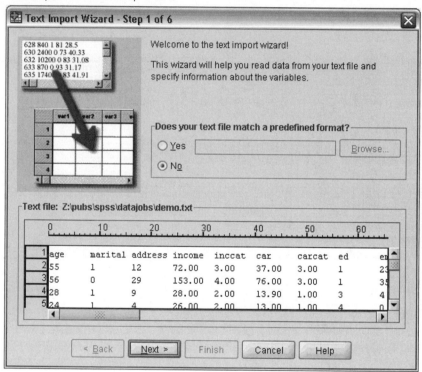

▶ In Step 1, you can choose a predefined format or create a new format in the wizard. Select No to indicate that a new format should be created.

▶ Click Next to continue.

As stated earlier, this file uses tab-delimited formatting. Also, the variable names are defined on the top line of this file.

Figure 2-13
Text Import Wizard: Step 2 of 6

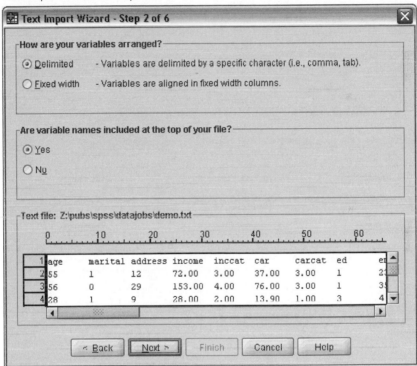

▶ Select Delimited to indicate that the data use a delimited formatting structure.

▶ Select Yes to indicate that variable names should be read from the top of the file.

▶ Click Next to continue.

▶ Type 2 in the top section of next dialog box to indicate that the first row of data starts on the second line of the text file.

Figure 2-14
Text Import Wizard: Step 3 of 6

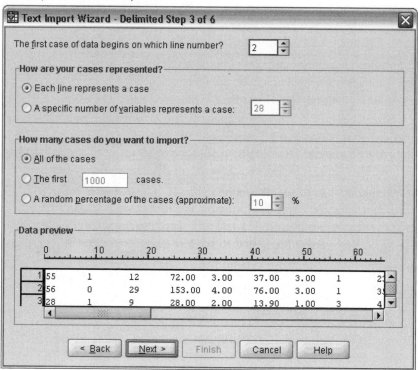

▶ Keep the default values for the remainder of this dialog box, and click Next to continue.

The Data preview in Step 4 provides you with a quick way to ensure that your data are being properly read.

Figure 2-15
Text Import Wizard: Step 4 of 6

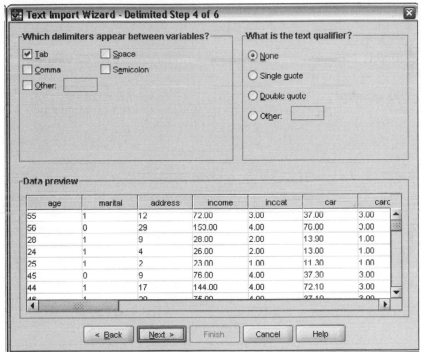

▶ Select Tab and deselect the other options.

▶ Click Next to continue.

Because the variable names may have been truncated to fit formatting requirements, this dialog box gives you the opportunity to edit any undesirable names.

Figure 2-16
Text Import Wizard: Step 5 of 6

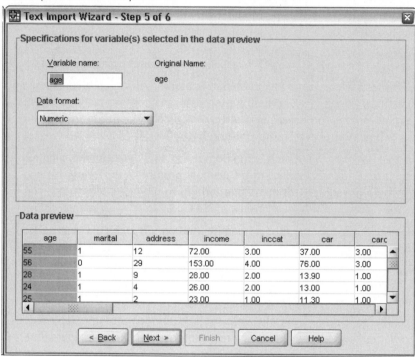

Data types can be defined here as well. For example, it's safe to assume that the income variable is meant to contain a certain dollar amount.

To change a data type:

▶ Under Data preview, select the variable you want to change, which is *Income* in this case.

► Select Dollar from the Data format drop-down list.

Figure 2-17
Change the data type

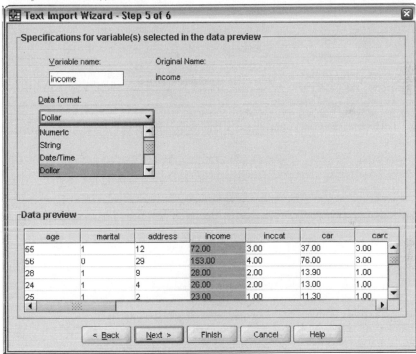

► Click Next to continue.

Figure 2-18
Text Import Wizard: Step 6 of 6

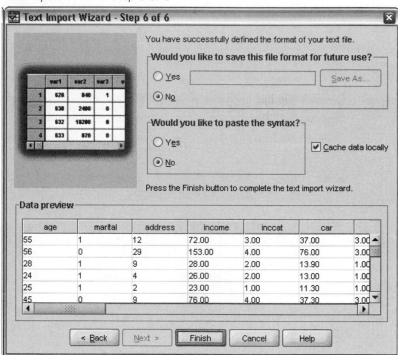

▶ Leave the default selections in this dialog box, and click Finish to import the data.

Saving Data

To save an SPSS-format data file, the Data Editor window must be the active window.

▶ From the menus choose:
File
  Save

▶ Go to the desired directory.

▶ Type a name for the file.

The Variables button can be used to select which variables in the Data Editor are saved to the SPSS-format data file. By default, all variables in the Data Editor are retained.

▶ Click Save.

The name in the title bar of the Data Editor will change to the filename you specified. This confirms that the file has been successfully saved as an SPSS-format data file. The file contains both variable information (names, type, and, if provided, labels and missing value codes) and all data values.

Using the Data Editor

The Data Editor displays the contents of the active data file. The information in the Data Editor consists of variables and cases.

- In Data View, columns represent variables, and rows represent cases (observations).
- In Variable View, each row is a variable, and each column is an attribute that is associated with that variable.

Variables are used to represent the different types of data that you have compiled. A common analogy is that of a survey. The response to each question on a survey is equivalent to a variable. Variables come in many different types, including numbers, strings, currency, and dates.

Entering Numeric Data

Data can be entered into the Data Editor, which may be useful for small data files or for making minor edits to larger data files.

▶ Click the Variable View tab at the bottom of the Data Editor window.

You need to define the variables that will be used. In this case, only three variables are needed: *age, marital status,* and *income.*

Figure 3-1
Variable names in Variable View

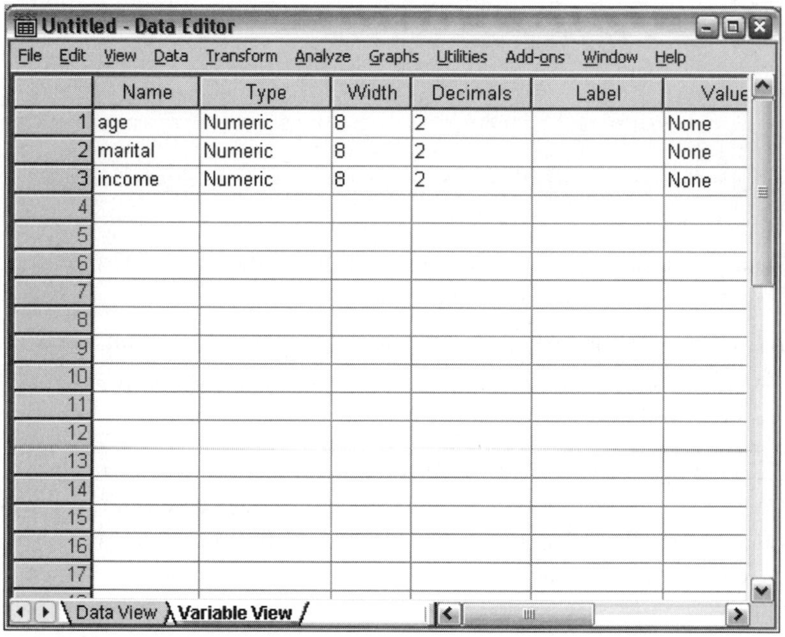

	Name	Type	Width	Decimals	Label	Value
1	age	Numeric	8	2		None
2	marital	Numeric	8	2		None
3	income	Numeric	8	2		None
4						
5						
6						
7						
8						
9						
10						
11						
12						
13						
14						
15						
16						
17						

▶ In the first row of the first column, type age.

▶ In the second row, type marital.

▶ In the third row, type income.

New variables are automatically given a Numeric data type.

If you don't enter variable names, unique names are automatically created. However, these names are not descriptive and are not recommended for large data files.

▶ Click the Data View tab to continue entering the data.

The names that you entered in Variable View are now the headings for the first three columns in Data View.

Begin entering data in the first row, starting at the first column.

Figure 3-2
Values entered in Data View

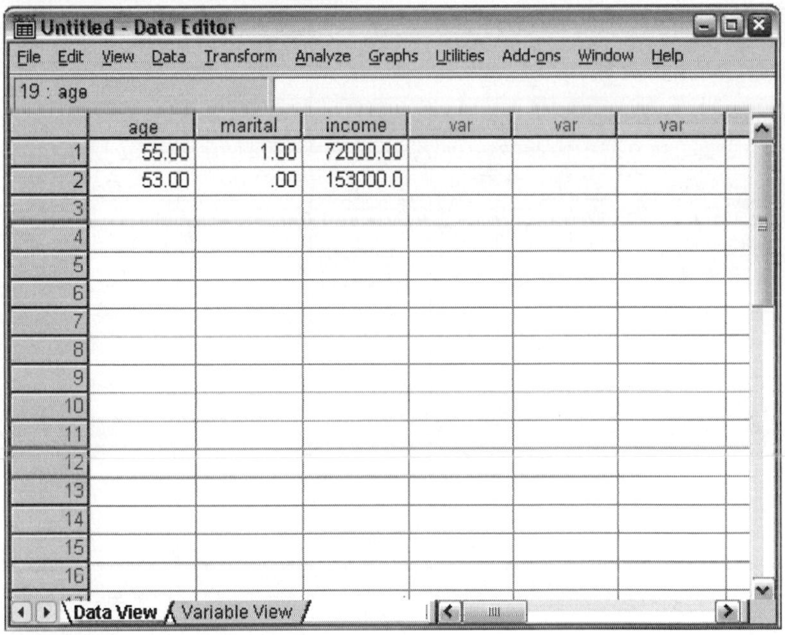

▶ In the *age* column, type 55.

▶ In the *marital* column, type 1.

▶ In the *income* column, type 72000.

▶ Move the cursor to the second row of the first column to add the next subject's data.

▶ In the *age* column, type 53.

▶ In the *marital* column, type 0.

▶ In the *income* column, type 153000.

Currently, the *age* and *marital* columns display decimal points, even though their values are intended to be integers. To hide the decimal points in these variables:

▶ Click the Variable View tab at the bottom of the Data Editor window.

▶ In the *Decimals* column of the *age* row, type 0 to hide the decimal.

▶ In the *Decimals* column of the *marital* row, type 0 to hide the decimal.

Figure 3-3
Updated decimal property for age and marital

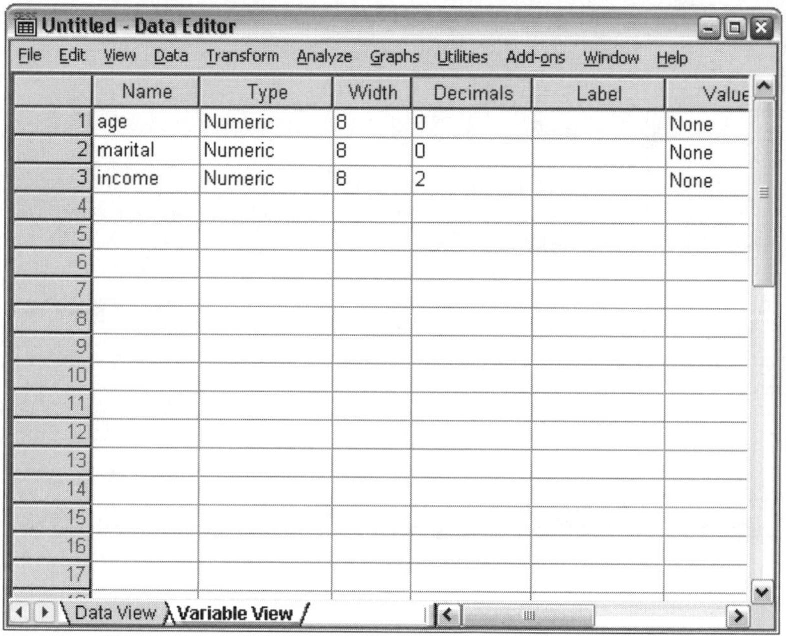

		Name	Type	Width	Decimals	Label	Value
1	age	Numeric	8	0		None	
2	marital	Numeric	8	0		None	
3	income	Numeric	8	2		None	
4							
5							
6							
7							
8							
9							
10							
11							
12							
13							
14							
15							
16							
17							

Data View / Variable View

Entering String Data

Non-numeric data, such as strings of text, can also be entered into the Data Editor.

▶ Click the Variable View tab at the bottom of the Data Editor window.

▶ In the first cell of the first empty row, type **sex** for the variable name.

▶ Click the *Type* cell next to your entry.

▶ Click the button on the right side of the *Type* cell to open the Variable Type dialog box.

Figure 3-4
Button shown in Type cell for sex

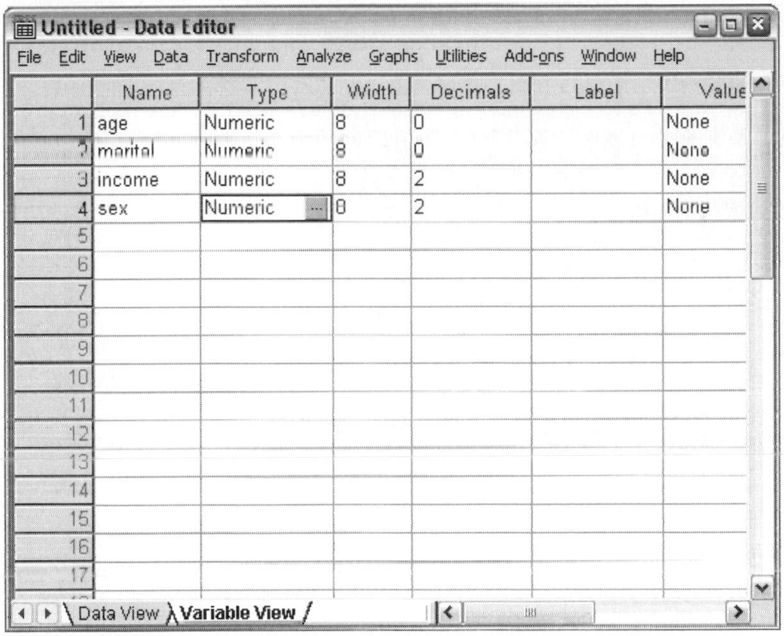

	Name	Type	Width	Decimals	Label	Value
1	age	Numeric	8	0		None
2	marital	Numeric	8	0		None
3	income	Numeric	8	2		None
4	sex	Numeric ...	8	2		None
5						
6						
7						
8						
9						
10						
11						
12						
13						
14						
15						
16						
17						

Untitled - Data Editor

File Edit View Data Transform Analyze Graphs Utilities Add-ons Window Help

◄ ► \ Data View \ Variable View /

▶ Select String to specify the variable type.

▶ Click OK to save your selection and return to the Data Editor.

Figure 3-5
Variable Type dialog box

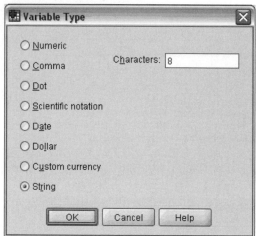

Defining Data

In addition to defining data types, you can also define descriptive variable labels and value labels for variable names and data values. These descriptive labels are used in statistical reports and charts.

Adding Variable Labels

Labels are meant to provide descriptions of variables. These descriptions are often longer versions of variable names. Labels can be up to 255 bytes. These labels are used in your output to identify the different variables.

▶ Click the Variable View tab at the bottom of the Data Editor window.

▶ In the *Label* column of the *age* row, type Respondent's Age.

▶ In the *Label* column of the *marital* row, type Marital Status.

▶ In the *Label* column of the *income* row, type Household Income.

▶ In the *Label* column of the *sex* row, type Gender.

Figure 3-6
Variable labels entered in Variable View

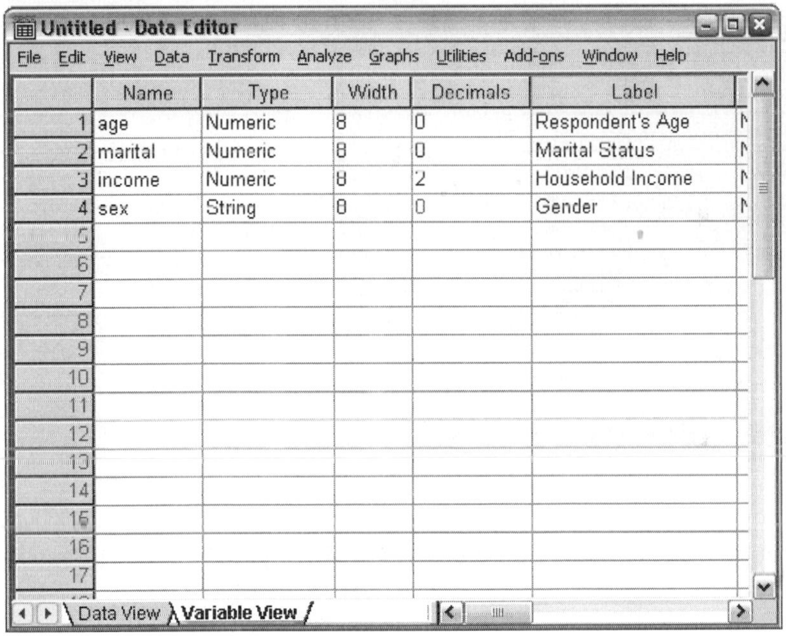

	Name	Type	Width	Decimals	Label	
1	age	Numeric	8	0	Respondent's Age	N
2	marital	Numeric	8	0	Marital Status	N
3	income	Numeric	8	2	Household Income	N
4	sex	String	8	0	Gender	N
5						
6						
7						
8						
9						
10						
11						
12						
13						
14						
15						
16						
17						

Data View \ Variable View

Changing Variable Type and Format

The *Type* column displays the current data type for each variable. The most common data types are numeric and string, but many other formats are supported. In the current data file, the *income* variable is defined as a numeric type.

▶ Click the *Type* cell for the *income* row, and then click the button on the right side of the cell to open the Variable Type dialog box.

▶ Select Dollar.

Figure 3-7
Variable Type dialog box

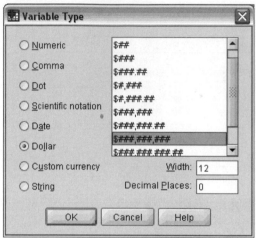

The formatting options for the currently selected data type are displayed.

▶ For the format of the currency in this example, select $###,###,###.

▶ Click OK to save your changes.

Adding Value Labels for Numeric Variables

Value labels provide a method for mapping your variable values to a string label. In this example, there are two acceptable values for the *marital* variable. A value of 0 means that the subject is single, and a value of 1 means that he or she is married.

▶ Click the *Values* cell for the *marital* row, and then click the button on the right side of the cell to open the Value Labels dialog box.

The **value** is the actual numeric value.

The **value label** is the string label that is applied to the specified numeric value.

▶ Type 0 in the Value field.

▶ Type Single in the Label field.

▶ Click Add to add this label to the list.

Figure 3-8
Value Labels dialog box

▶ Type 1 in the Value field, and type Married in the Label field.

▶ Click Add, and then click OK to save your changes and return to the Data Editor.

These labels can also be displayed in Data View, which can make your data more readable.

▶ Click the Data View tab at the bottom of the Data Editor window.

▶ From the menus choose:
View
  Value Labels

The labels are now displayed in a list when you enter values in the Data Editor. This setup has the benefit of suggesting a valid response and providing a more descriptive answer.

If the Value Labels menu item is already active (with a check mark next to it), choosing Value Labels again will turn *off* the display of value labels.

Figure 3-9
Value labels displayed in Data View

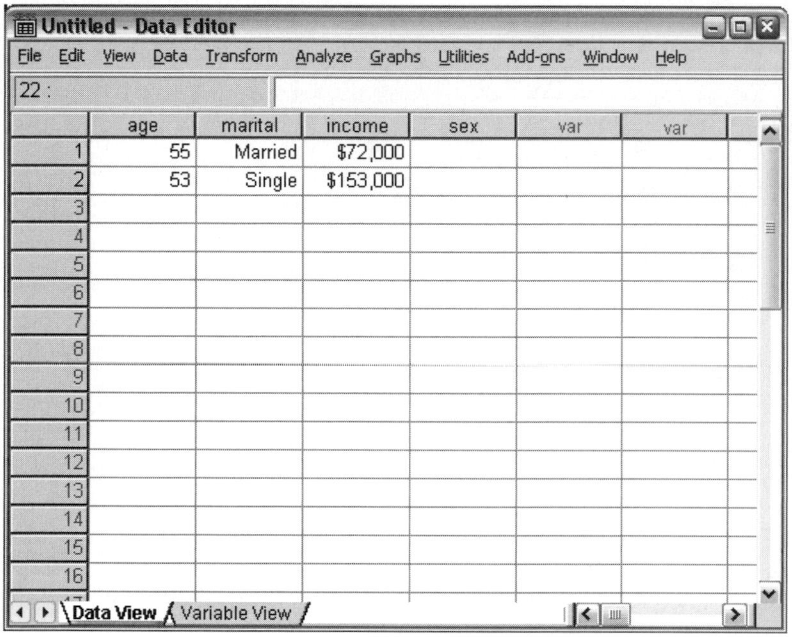

Adding Value Labels for String Variables

String variables may require value labels as well. For example, your data may use single letters, *M* or *F*, to identify the sex of the subject. Value labels can be used to specify that *M* stands for *Male* and *F* stands for *Female*.

▶ Click the Variable View tab at the bottom of the Data Editor window.

▶ Click the *Values* cell in the *sex* row, and then click the button on the right side of the cell to open the Value Labels dialog box.

▶ Type F in the Value field, and then type Female in the Label field.

▶ Click Add to add this label to your data file.

Figure 3-10
Value Labels dialog box

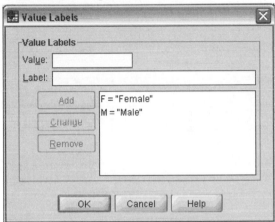

▶ Type M in the Value field, and type Male in the Label field.

▶ Click Add, and then click OK to save your changes and return to the Data Editor.

Because string values are case sensitive, you should be consistent. A lowercase *m* is not the same as an uppercase *M*.

Using Value Labels for Data Entry

You can use value labels for data entry.

▶ Click the Data View tab at the bottom of the Data Editor window.

▶ In the first row, select the cell for *sex*.

▶ Click the button on the right side of the cell, and then choose Male from the drop-down list.

▶ In the second row, select the cell for *sex*.

▶ Click the button on the right side of the cell, and then choose Female from the drop-down list.

Figure 3-11
Using variable labels to select values

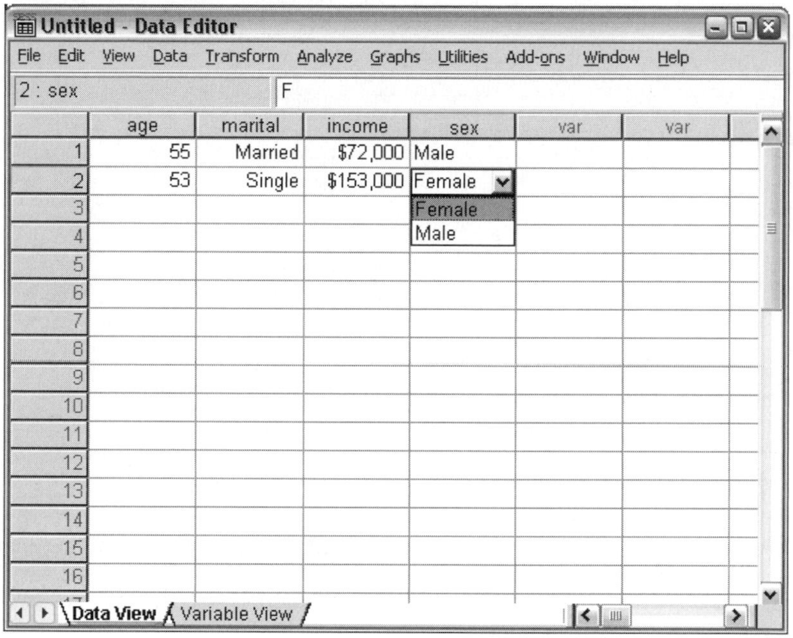

Only defined values are listed, which ensures that the entered data are in a format that you expect.

Handling Missing Data

Missing or invalid data are generally too common to ignore. Survey respondents may refuse to answer certain questions, may not know the answer, or may answer in an unexpected format. If you don't filter or identify these data, your analysis may not provide accurate results.

For numeric data, empty data fields or fields containing invalid entries are converted to system-missing, which is identifiable by a single period.

Figure 3-12
Missing values displayed as periods

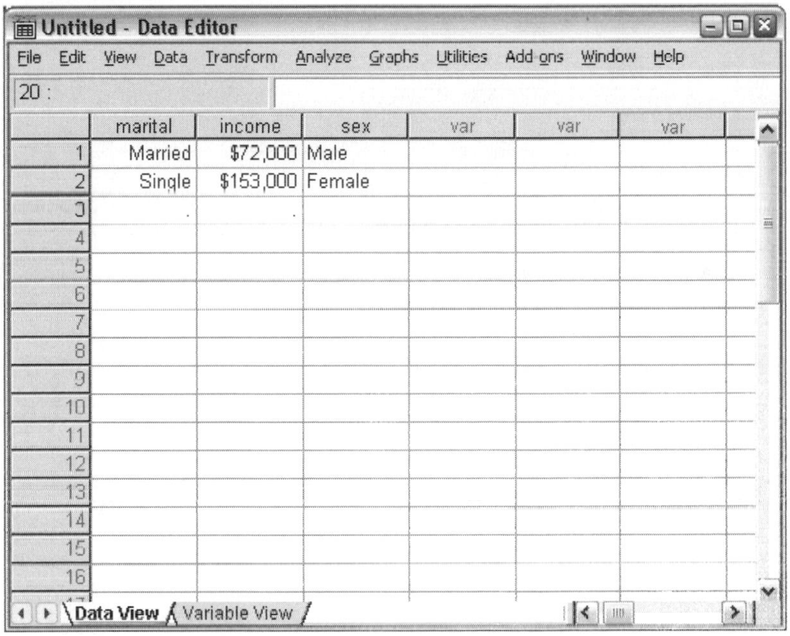

The reason a value is missing may be important to your analysis. For example, you may find it useful to distinguish between those respondents who refused to answer a question and those respondents who didn't answer a question because it was not applicable.

Missing Values for a Numeric Variable

▶ Click the Variable View tab at the bottom of the Data Editor window.

▶ Click the *Missing* cell in the *age* row, and then click the button on the right side of the cell to open the Missing Values dialog box.

In this dialog box, you can specify up to three distinct missing values, or you can specify a range of values plus one additional discrete value.

Figure 3-13
Missing Values dialog box

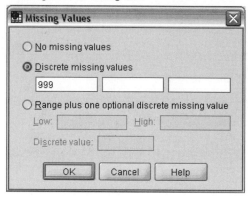

▶ Select Discrete missing values.

▶ Type 999 in the first text box and leave the other two text boxes empty.

▶ Click OK to save your changes and return to the Data Editor.

Now that the missing data value has been added, a label can be applied to that value.

▶ Click the *Values* cell in the *age* row, and then click the button on the right side of the cell to open the Value Labels dialog box.

▶ Type 999 in the Value field.

▶ Type No Response in the Label field.

Figure 3-14
Value Labels dialog box

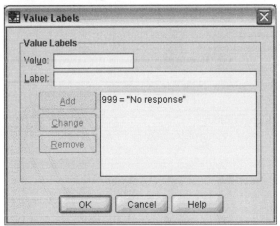

▶ Click Add to add this label to your data file.

▶ Click OK to save your changes and return to the Data Editor.

Missing Values for a String Variable

Missing values for string variables are handled similarly to the missing values for numeric variables. However, unlike numeric variables, empty fields in string variables are not designated as system-missing. Rather, they are interpreted as an empty string.

▶ Click the Variable View tab at the bottom of the Data Editor window.

▶ Click the *Missing* cell in the *sex* row, and then click the button on the right side of the cell to open the Missing Values dialog box.

▶ Select Discrete missing values.

▶ Type NR in the first text box.

Missing values for string variables are case sensitive. So, a value of *nr* is not treated as a missing value.

▶ Click OK to save your changes and return to the Data Editor.

Now you can add a label for the missing value.

▶ Click the *Values* cell in the *sex* row, and then click the button on the right side of the cell to open the Value Labels dialog box.

▶ Type NR in the Value field.

▶ Type No Response in the Label field.

Figure 3-15
Value Labels dialog box

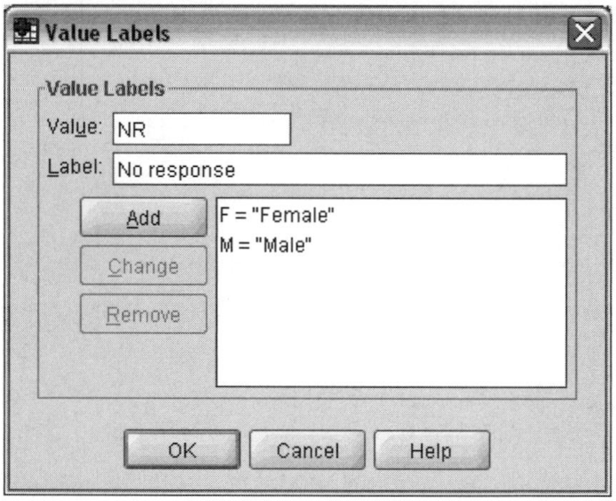

▶ Click Add to add this label to your project.

▶ Click OK to save your changes and return to the Data Editor.

Copying and Pasting Variable Attributes

After you've defined variable attributes for a variable, you can copy these attributes and apply them to other variables.

▶ In Variable View, type **agewed** in the first cell of the first empty row.

Figure 3-16
agewed variable in Variable View

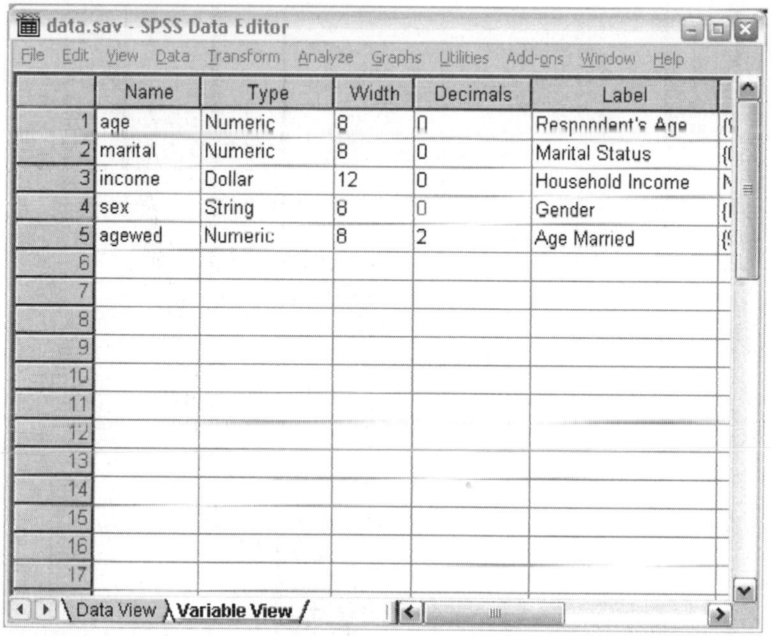

	Name	Type	Width	Decimals	Label	
1	age	Numeric	8	0	Respondent's Age	{(
2	marital	Numeric	8	0	Marital Status	{(
3	income	Dollar	12	0	Household Income	N
4	sex	String	8	0	Gender	{(
5	agewed	Numeric	8	2	Age Married	{(
6						
7						
8						
9						
10						
11						
12						
13						
14						
15						
16						
17						

data.sav - SPSS Data Editor

File Edit View Data Transform Analyze Graphs Utilities Add-ons Window Help

Data View \ Variable View /

▶ In the *Label* column, type **Age Married**.

▶ Click the *Values* cell in the *age* row.

▶ From the menus choose:
Edit
  Copy

▶ Click the *Values* cell in the *agewed* row.

▶ From the menus choose:
Edit
  Paste

The defined values from the *age* variable are now applied to the *agewed* variable.

To apply the attribute to multiple variables, simply select multiple target cells (click and drag down the column).

Figure 3-17
Multiple cells selected

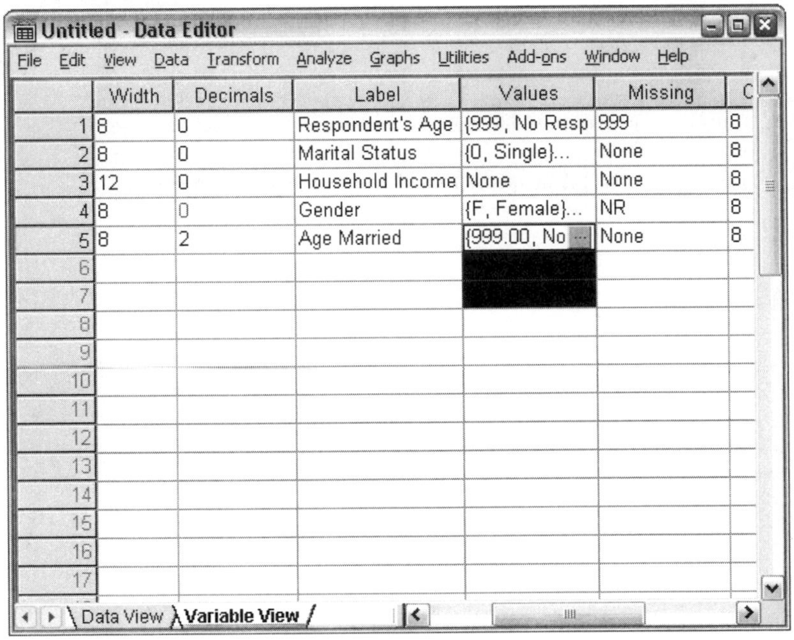

	Width	Decimals	Label	Values	Missing	C
1	8	0	Respondent's Age	{999, No Resp	999	8
2	8	0	Marital Status	{0, Single}...	None	8
3	12	0	Household Income	None	None	8
4	8	0	Gender	{F, Female}...	NR	8
5	8	2	Age Married	{999.00, No ⋯	None	8
6						
7						
8						
9						
10						
11						
12						
13						
14						
15						
16						
17						

When you paste the attribute, it is applied to all of the selected cells.

New variables are automatically created if you paste the values into empty rows.

To copy all attributes from one variable to another variable:

▶ Click the row number in the *marital* row.

Figure 3-18
Selected row

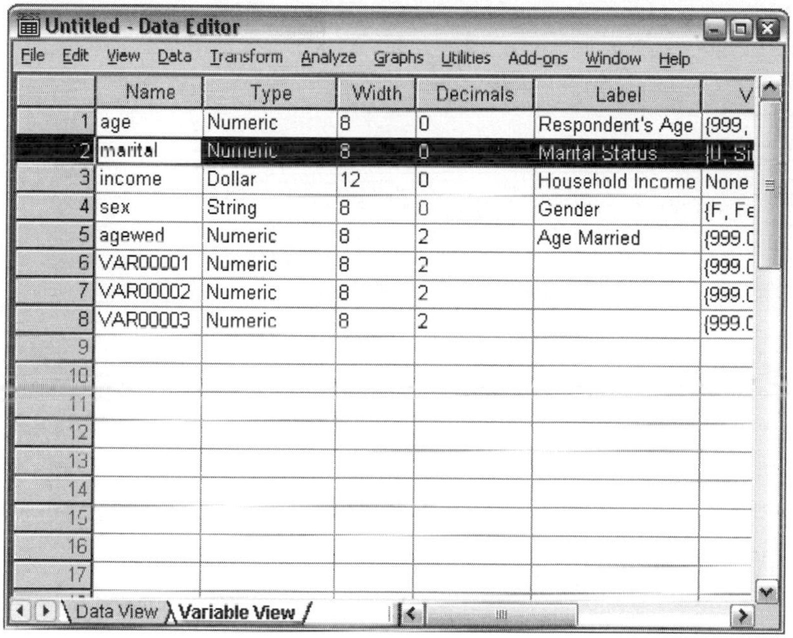

▶ From the menus choose:
Edit
  Copy

▶ Click the row number of the first empty row.

▶ From the menus choose:
Edit
  Paste

All attributes of the *marital* variable are applied to the new variable.

Figure 3-19
All values pasted into a row

Defining Variable Properties for Categorical Variables

For categorical (nominal, ordinal) data, you can use Define Variable Properties to define value labels and other variable properties. The Define Variable Properties process:

- Scans the actual data values and lists all unique data values for each selected variable.
- Identifies unlabeled values and provides an "auto-label" feature.
- Provides the ability to copy defined value labels from another variable to the selected variable or from the selected variable to additional variables.

This example uses the data file *demo.sav*. This data file already has defined value labels, so we will enter a value for which there is no defined value label.

▶ In Data View of the Data Editor, click the first data cell for the variable *ownpc* (you may have to scroll to the right), and then enter 99.

▶ From the menus choose:

Data
  Define Variable Properties...

Figure 3-20
Initial Define Variable Properties dialog box

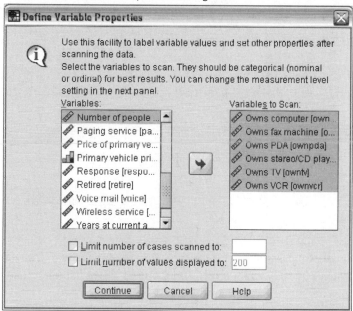

In the initial Define Variable Properties dialog box, you select the nominal or ordinal variables for which you want to define value labels and/or other properties.

▶ Drag and drop *Owns computer [ownpc]* through *Owns VCR [ownvcr]* into the Variables to Scan list.

You might notice that the measurement level icons for all of the selected variables indicate that they are scale variables, not categorical variables. All of the selected variables in this example are really categorical variables that use the numeric values 0 and 1 to stand for *No* and *Yes*, respectively—and one of the variable properties that we'll change with Define Variable Properties is the measurement level.

▶ Click Continue.

Figure 3-21
Define Variable Properties main dialog box

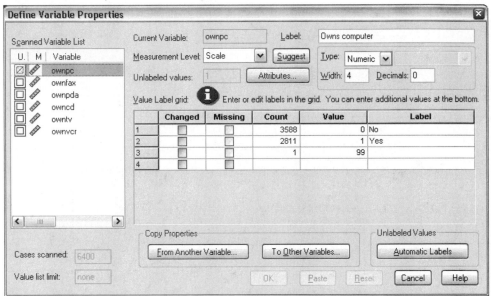

▶ In the Scanned Variable List, select *ownpc*.

The current level of measurement for the selected variable is scale. You can change the measurement level by selecting a level from the drop-down list, or you can let Define Variable Properties suggest a measurement level.

▶ Click Suggest.

The Suggest Measurement Level dialog box is displayed.

Figure 3-22
Suggest Measurement Level dialog box

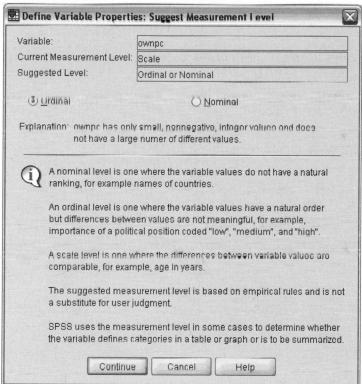

Because the variable doesn't have very many different values and all of the scanned cases contain integer values, the proper measurement level is probably ordinal or nominal.

▶ Select Ordinal, and then click Continue.

The measurement level for the selected variable is now ordinal.

The Value Label grid displays all of the unique data values for the selected variable, any defined value labels for these values, and the number of times (count) that each value occurs in the scanned cases.

The value that we entered in Data View, 99, is displayed in the grid. The count is only 1 because we changed the value for only one case, and the *Label* column is empty because we haven't defined a value label for 99 yet. An X in the first column of the

Scanned Variable List also indicates that the selected variable has at least one observed value without a defined value label.

▶ In the *Label* column for the value of 99, enter No answer.

▶ Check the box in the *Missing* column for the value 99 to identify the value 99 as **user-missing**.
Data values that are specified as user-missing are flagged for special treatment and are excluded from most calculations.

Figure 3-23
New variable properties defined for ownpc

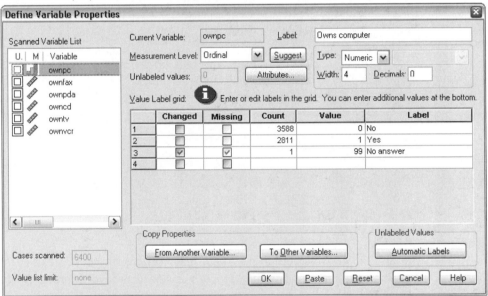

Before we complete the job of modifying the variable properties for *ownpc*, let's apply the same measurement level, value labels, and missing values definitions to the other variables in the list.

▶ In the Copy Properties area, click To Other Variables.

Figure 3-24
Apply Labels and Level dialog box

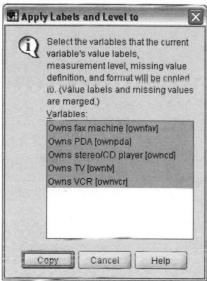

▶ In the Apply Labels and Level dialog box, select all of the variables in the list, and then click Copy.

If you select any other variable in the Scanned Variable List of the Define Variable Properties main dialog box now, you'll see that they are all ordinal variables, with a value of 99 defined as user-missing and a value label of *No answer*.

Figure 3-25
New variable properties defined for ownfax

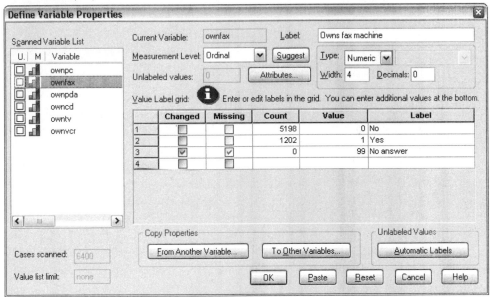

▶ Click OK to save all of the variable properties that you have defined.

Working with Multiple Data Sources

Starting with version 14.0, multiple data sources can be open at the same time, making it easier to:

- Switch back and forth between data sources.

- Compare the contents of different data sources.

- Copy and paste data between data sources.

- Create multiple subsets of cases and/or variables for analysis.

- Merge multiple data sources from various data formats (for example, spreadsheet, database, text data) without saving each data source first.

Basic Handling of Multiple Data Sources

Figure 4-1
Two data sources open at same time

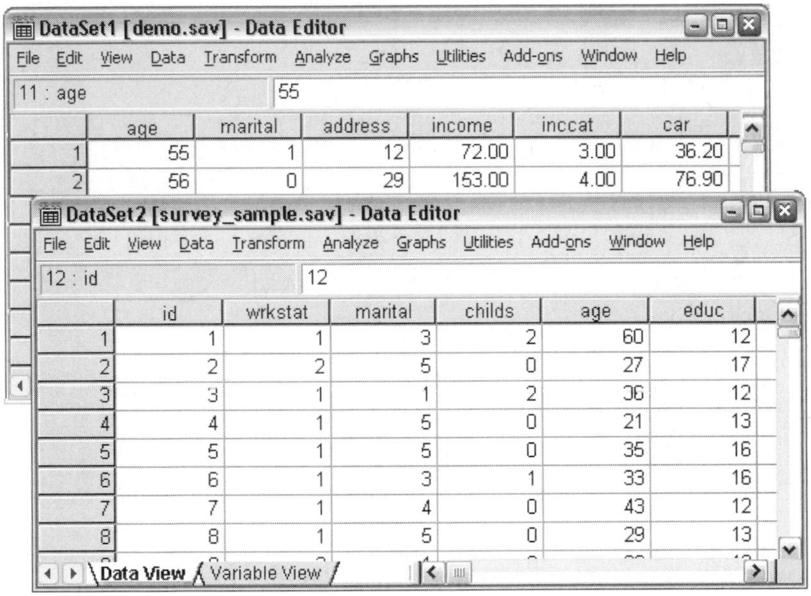

By default, each data source that you open is displayed in a new Data Editor window.

- Any previously open data sources remain open and available for further use.

- When you first open a data source, it automatically becomes the **active dataset**.

- You can change the active dataset simply by clicking anywhere in the Data Editor window of the data source that you want to use or by selecting the Data Editor window for that data source from the Window menu.

- Only the variables in the active dataset are available for analysis.

Figure 4-2
Variable list containing variables in the active dataset

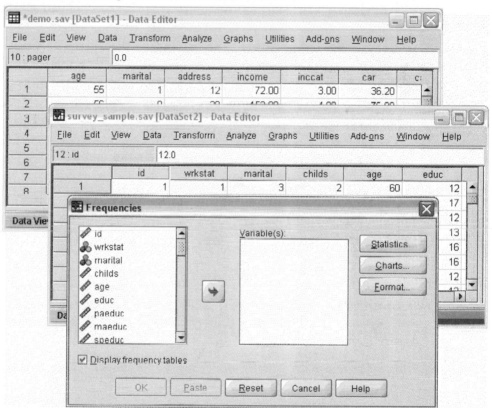

- You cannot change the active dataset when any dialog box that accesses the data is open (including all dialog boxes that display variable lists).

- At least one Data Editor window must be open during a session. When you close the last open Data Editor window, SPSS automatically shuts down, prompting you to save changes first.

Copying and Pasting Information between Datasets

You can copy both data and variable definition attributes from one dataset to another dataset in basically the same way that you copy and paste information within a single data file.

- Copying and pasting selected data cells in Data View pastes only the data values, with no variable definition attributes.

- Copying and pasting an entire variable in Data View by selecting the variable name at the top of the column pastes all of the data and all of the variable definition attributes for that variable.

- Copying and pasting variable definition attributes or entire variables in Variable View pastes the selected attributes (or the entire variable definition) but does not paste any data values.

Renaming Datasets

When you open a data source through the menus and dialog boxes, each data source is automatically assigned a dataset name of *DataSetn*, where *n* is a sequential integer value, and when you open a data source using command syntax, no dataset name is assigned unless you explicitly specify one with DATASET NAME. To provide more descriptive dataset names:

▶ From the menus in the Data Editor window for the dataset whose name you want to change choose:

File
  Rename Dataset...

▶ Enter a new dataset name that conforms to variable naming rules.

Suppressing Multiple Datasets

If you prefer to have only one dataset available at a time and want to suppress the multiple dataset feature:

▶ From the menus choose:

Edit
  Options...

▶ Click the General tab.

Select (check) Open only one dataset at a time.

Examining Summary Statistics for Individual Variables

This chapter discusses simple summary measures and how the level of measurement of a variable influences the types of statistics that should be used. We will use the data file *demo.sav*. For more information, see "Sample Files" in Appendix A on p. 201.

Level of Measurement

Different summary measures are appropriate for different types of data, depending on the level of measurement:

Categorical. Data with a limited number of distinct values or categories (for example, gender or marital status). Also referred to as **qualitative data**. Categorical variables can be string (alphanumeric) data or numeric variables that use numeric codes to represent categories (for example, $0 = Unmarried$ and $1 = Married$). There are two basic types of categorical data:

- **Nominal**. Categorical data where there is no inherent order to the categories. For example, a job category of *sales* is not higher or lower than a job category of *marketing* or *research*.

- **Ordinal**. Categorical data where there is a meaningful order of categories, but there is not a measurable distance between categories. For example, there is an order to the values *high, medium,* and *low*, but the "distance" between the values cannot be calculated.

Scale. Data measured on an interval or ratio scale, where the data values indicate both the order of values and the distance between values. For example, a salary of $72,195 is higher than a salary of $52,398, and the distance between the two values is $19,797. Also referred to as **quantitative** or **continuous data**.

Summary Measures for Categorical Data

For categorical data, the most typical summary measure is the number or percentage of cases in each category. The **mode** is the category with the greatest number of cases. For ordinal data, the **median** (the value at which half of the cases fall above and below) may also be a useful summary measure if there is a large number of categories.

The Frequencies procedure produces frequency tables that display both the number and percentage of cases for each observed value of a variable.

▶ From the menus choose:
Analyze
  Descriptive Statistics
    Frequencies...

▶ Select *Owns PDA [ownpda]* and *Owns TV [owntv]* and move them into the Variable(s) list.

Figure 5-1
Categorical variables selected for analysis

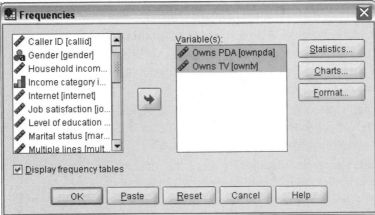

▶ Click OK to run the procedure.

Figure 5-2
Frequency tables

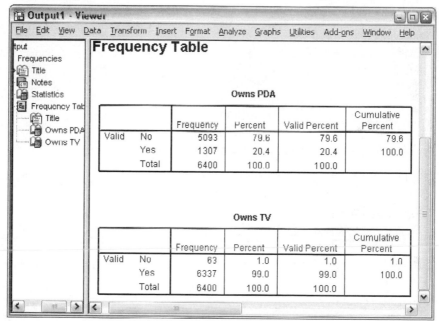

The frequency tables are displayed in the Viewer window. The frequency tables reveal that only 20.4% of the people own PDAs, but almost everybody owns a TV (99.0%). These might not be interesting revelations, although it might be interesting to find out more about the small group of people who do not own televisions.

Charts for Categorical Data

You can graphically display the information in a frequency table with a bar chart or pie chart.

▶ Open the Frequencies dialog box again. (The two variables should still be selected.)

You can use the Dialog Recall button on the toolbar to quickly return to recently used procedures.

Figure 5-3
Dialog Recall button

▶ Click Charts.

▶ Select **Bar charts** and then click Continue.

Figure 5-4
Frequencies Charts dialog box

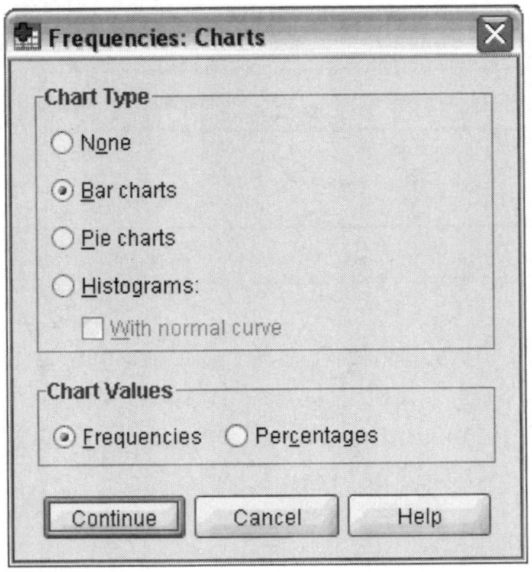

▶ Click OK in the main dialog box to run the procedure.

Figure 5-5
Bar chart

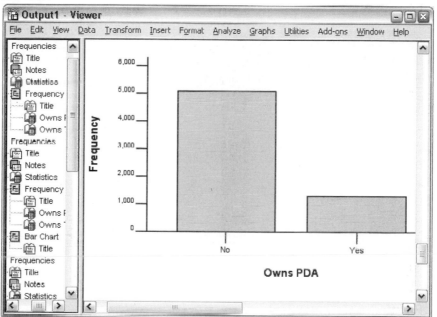

In addition to the frequency tables, the same information is now displayed in the form of bar charts, making it easy to see that most people do not own PDAs but almost everyone owns a TV.

Summary Measures for Scale Variables

There are many summary measures available for scale variables, including:

- **Measures of central tendency.** The most common measures of central tendency are the **mean** (arithmetic average) and **median** (value at which half the cases fall above and below).

- **Measures of dispersion.** Statistics that measure the amount of variation or spread in the data include the standard deviation, minimum, and maximum.

▶ Open the Frequencies dialog box again.

▶ Click Reset to clear any previous settings.

▶ Select *Household income in thousands [income]* and move it into the Variable(s) list.

Figure 5-6
Scale variable selected for analysis

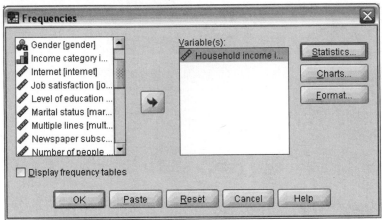

▶ Click Statistics.

▶ Select Mean, Median, Std. deviation, Minimum, and Maximum.

Figure 5-7
Frequencies Statistics dialog box

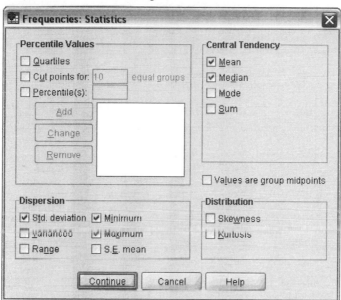

▶ Click Continue.

▶ Deselect Display frequency tables in the main dialog box. (Frequency tables are usually not useful for scale variables since there may be almost as many distinct values as there are cases in the data file.)

▶ Click OK to run the procedure.

The Frequencies Statistics table is displayed in the Viewer window.

Figure 5-8
Frequencies Statistics table

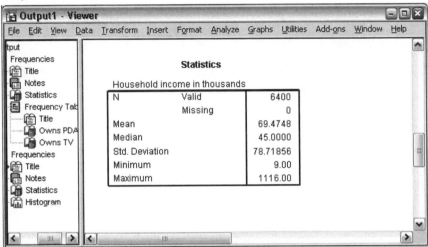

In this example, there is a large difference between the mean and the median. The mean is almost 25,000 greater than the median, indicating that the values are not normally distributed. You can visually check the distribution with a histogram.

Histograms for Scale Variables

▶ Open the Frequencies dialog box again.

▶ Click Charts.

▶ Select Histograms and With normal curve.

Figure 5-9
Frequencies Charts dialog box

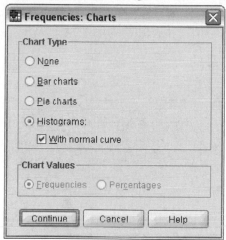

▶ Click Continue, and then click OK in the main dialog box to run the procedure.

Figure 5-10
Histogram

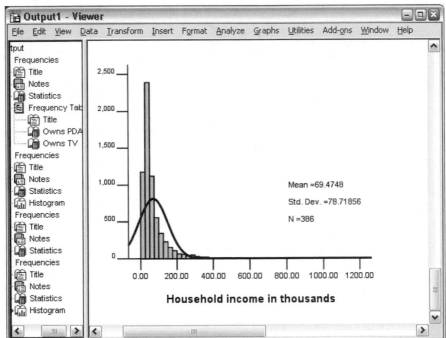

The majority of cases are clustered at the lower end of the scale, with most falling below 100,000. There are, however, a few cases in the 500,000 range and beyond (too few to even be visible without modifying the histogram). These high values for only a few cases have a significant effect on the mean but little or no effect on the median, making the median a better indicator of central tendency in this example.

Creating and Editing Charts

You can create and edit a wide variety of chart types. In this chapter, we will create and edit bar charts. You can apply the principles to any chart type.

Chart Creation Basics

To demonstrate the basics of chart creation, we will create a bar chart of mean income for different levels of job satisfaction. This example uses the data file *demo.sav*. For more information, see "Sample Files" in Appendix A on p. 201.

▶ From the menus choose:
Graphs
  Chart Builder...

The Chart Builder dialog box is an interactive window that allows you to preview
how a chart will look while you build it.

Figure 6-1
Chart Builder dialog box

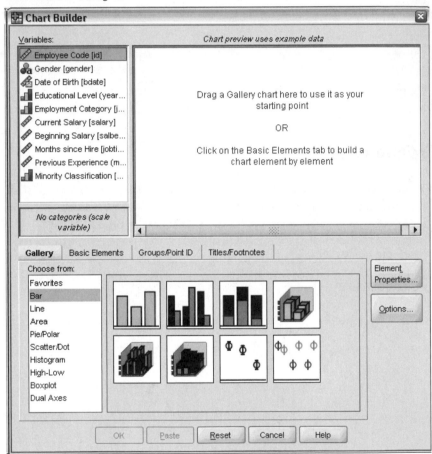

Using the Chart Builder Gallery

▶ Click the Gallery tab if it is not selected.

The Gallery includes many different predefined charts, which are organized by chart type. The Basic Elements tab also provides basic elements (such as axes and graphic elements) for creating charts from scratch, but it's easier to use the Gallery.

▶ Click Bar if it is not selected.

Icons representing the available bar charts in the Gallery appear in the dialog box. The pictures should provide enough information to identify the specific chart type. If you need more information, you can also display a ToolTip description of the chart by pausing your cursor over an icon.

▶ Drag the icon for the simple bar chart onto the "canvas," which is the large area above the Gallery. The Chart Builder displays a preview of the chart on the canvas. Note that the data used to draw the chart are not your actual data. They are example data.

Figure 6-2
Bar chart on Chart Builder canvas

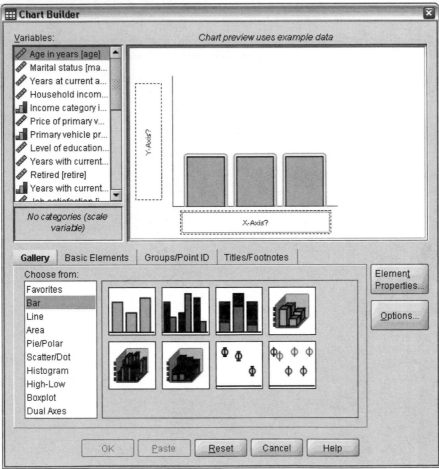

Defining Variables and Statistics

Although there is a chart on the canvas, it is not complete because there are no variables or statistics to control how tall the bars are and to specify which variable category corresponds to each bar. You can't have a chart without variables and statistics. You add variables by dragging them from the Variables list, which is located to the left of the canvas.

A variable's measurement level is important in the Chart Builder. You are going to use the *Job satisfaction* variable on the *x* axis. However, the icon (which looks like a ruler) next to the variable indicates that its measurement level is defined as scale. To create the correct chart, you must use a categorical measurement level. Instead of going back and changing the measurement level in the Variable View, you can change the measurement level temporarily in the Chart Builder.

▶ Right-click *Job satisfaction* in the Variables list and choose Ordinal. Ordinal is an appropriate measurement level because the categories in *Job satisfaction* can be ranked by level of satisfaction. Note that the icon changes after you change the measurement level.

▶ Now drag *Job satisfaction* from the Variables list to the *x* axis drop zone.

Figure 6-3
Job satisfaction in x axis drop zone

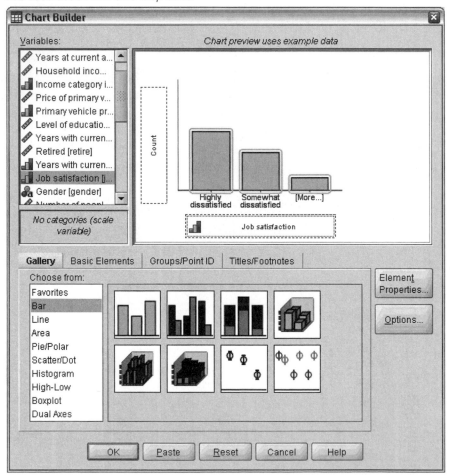

The *y* axis drop zone defaults to the *Count* statistic. If you want to use another statistic (such as percentage or mean), you can easily change it. You will not use either of these statistics in this example, but we will review the process in case you need to change this statistic at another time.

▶ Click Element Properties to display the Element Properties window.

Figure 6-4
Element Properties window

The Element Properties window allows you to change the properties of the various chart elements. These elements include the graphic elements (such as the bars in the bar chart) and the axes on the chart. Select one of the elements in the Edit Properties of list to change the properties associated with that element. Also note the red X located to the right of the list. This button deletes a graphic element from the canvas. Because Bar1 is selected, the properties shown apply to graphic elements, specifically the bar graphic element.

The Statistic drop-down list shows the specific statistics that are available. The same statistics are usually available for every chart type. Be aware that some statistics require that the *y* axis drop zone contains a variable.

▶ Return to the Chart Builder dialog box and drag *Household income in thousands* from the Variables list to the *y* axis drop zone. Because the variable on the *y* axis is scalar and the *x* axis variable is categorical (ordinal is a type of categorical measurement level), the *y* axis drop zone defaults to the *Mean* statistic. These are the variables and statistics you want, so there is no need to change the element properties.

Adding Text

You can also add titles and footnotes to the chart.

▶ Click the Titles/Footnotes tab.

▶ Select Title 1.

Figure 6-5
Title 1 displayed on canvas

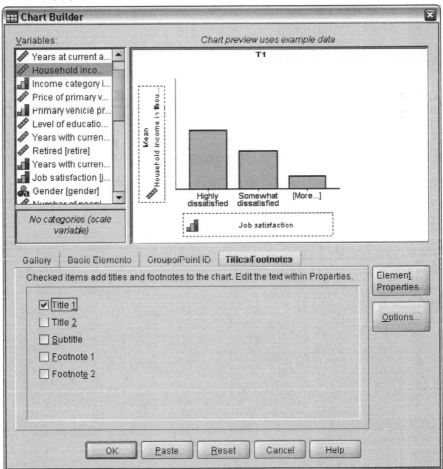

The title appears on the canvas with the label T1.

▶ In the Element Properties window, select Title 1 in the Edit Properties of list.

▶ In the Content text box, type Income by Job Satisfaction. This is the text that the title will display.

▶ Click Apply to save the text. Although the text is not displayed in the Chart Builder, it will appear when you generate the chart.

Creating the Chart

▶ Click OK to create the bar chart.

Figure 6-6
Bar chart

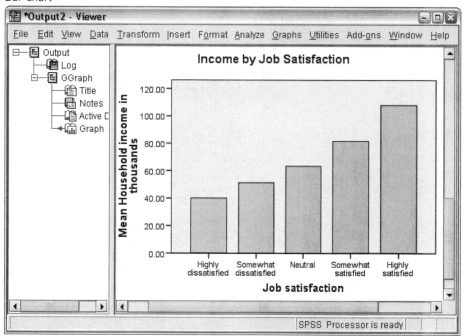

The bar chart reveals that respondents who are more satisfied with their jobs tend to have higher household incomes.

Chart Editing Basics

You can edit charts in a variety of ways. For the sample bar chart that you created, you will:

■ Change colors.

- Format numbers in tick labels.
- Edit text.
- Display data value labels.
- Use chart templates.

To edit the chart, open it in the Chart Editor.

▶ Double-click the bar chart to open it in the Chart Editor.

Figure 6-7
Bar chart in the Chart Editor

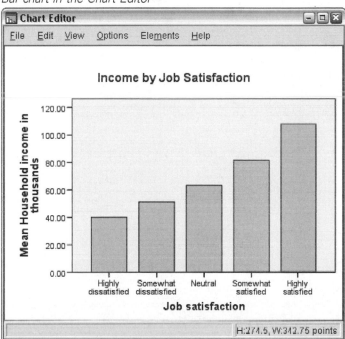

Selecting Chart Elements

To edit a chart element, you first select it.

▶ Click any one of the bars. The rectangles around the bars indicate that they are selected.

There are general rules for selecting elements in simple charts:

■ When no graphic elements are selected, click any graphic element to select all graphic elements.

■ When all graphic elements are selected, click a graphic element to select only that graphic element. You can select a different graphic element by clicking it. To select multiple graphic elements, click each element while pressing the Ctrl key.

▶ To deselect all elements, press the Esc key.

▶ Click any bar to select all of the bars again.

Using the Properties Window

▶ From the Chart Editor menus choose:
Edit
  Properties

This opens the Properties window, showing the tabs that apply to the bars you selected. These tabs change depending on what chart element you select in the Chart Editor. For example, if you had selected a text frame instead of bars, different tabs would appear in the Properties window. You will use these tabs to do most chart editing.

Figure 6-8
Properties window

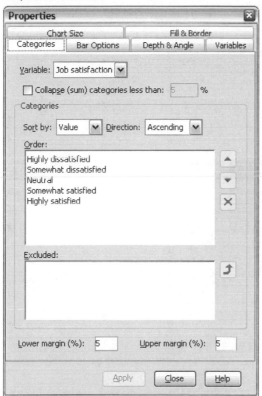

Changing Bar Colors

First, you will change the color of the bars. You specify color attributes of graphic elements (excluding lines and markers) on the Fill & Border tab.

▶ Click the Fill & Border tab.

▶ Click the swatch next to Fill to indicate that you want to change the fill color of the bars. The numbers below the swatch specify the red, green, and blue settings for the current color.

▶ Click the light blue color, which is second from the left in the second row from the bottom.

Figure 6-9
Fill & Border tab

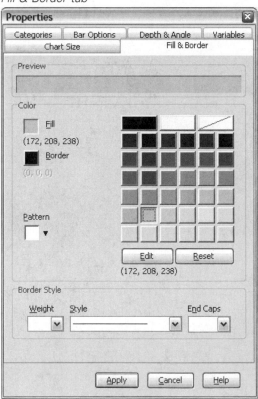

▶ Click Apply.

The bars in the chart are now light blue.

Figure 6-10
Edited bar chart showing blue bars

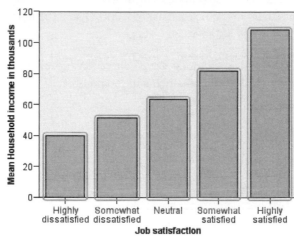

Formatting Numbers in Tick Labels

Notice that the numbers on the *y* axis are scaled in thousands. To make the chart more attractive and easier to interpret, we will change the number format in the tick labels and then edit the axis title appropriately.

▶ Select the *y* axis tick labels by clicking any one of them.

▶ To reopen the Properties window (if you closed it previously), from the menus choose:
Edit
  Properties

Note: From here on, we assume that the Properties window is open. If you have closed the Properties window, follow the previous step to reopen it. You can also use the keyboard shortcut Ctrl+T to reopen the window.

▶ Click the Number Format tab.

90

Chapter 6

▶ You do not want the tick labels to display decimal places, so type 0 in the Decimal Places text box.

▶ Type 0.001 in the Scaling Factor text box. The scaling factor is the number by which the Chart Editor divides the displayed number. Because 0.001 is a fraction, dividing by it will *increase* the numbers in the tick labels by 1,000. Thus, the numbers will no longer be in thousands; they will be unscaled.

▶ Select Display Digit Grouping. Digit grouping uses a character (specified by your computer's locale) to mark each thousandth place in the number.

Figure 6-11
Number Format tab

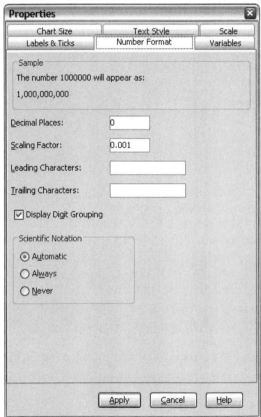

▶ Click Apply.

The tick labels reflect the new number formatting: There are no decimal places, the numbers are no longer scaled, and each thousandth place is specified with a character.

Figure 6-12
Edited bar chart showing new number format

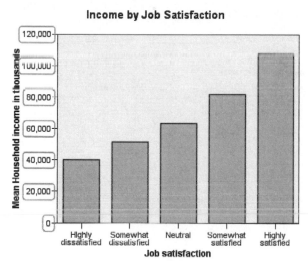

Editing Text

Now that you have changed the number format of the tick labels, the axis title is no longer accurate. Next, you will change the axis title to reflect the new number format.

Note: You do not need to open the Properties window to edit text. You can edit text directly on the chart.

▶ Click the *y* axis title to select it.

▶ Click the axis title again to start edit mode. While in edit mode, the Chart Editor positions any rotated text horizontally. It also displays a flashing red bar cursor (not shown in the example).

▶ Delete the following text:

in thousands

▶ Press Enter to exit edit mode and update the axis title. The axis title now accurately describes the contents of the tick labels.

Figure 6-13
Bar chart showing edited y axis title

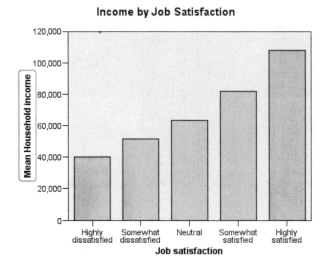

Displaying Data Value Labels

Another common task is to show the exact values associated with the graphic elements (which are bars in this example). These values are displayed in data labels.

▶ From the Chart Editor menus choose:
Elements
   Show Data Labels

Figure 6-14
Bar chart showing data value labels

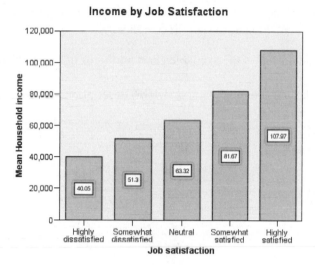

Each bar in the chart now displays the exact mean household income. Notice that the units are in thousands, so you could use the Number Format tab again to change the scaling factor.

Using Templates

If you make a number of routine changes to your charts, you can use a chart template to reduce the time needed to create and edit charts. A chart template saves the attributes of a specific chart. You can then apply the template when creating or editing a chart.

We will save the current chart as a template and then apply that template while creating a new chart.

► From the menus choose:
File
  Save Chart Template...

The Save Chart Template dialog box allows you to specify which chart attributes you want to include in the template.

If you expand any of the items in the tree view, you can see which specific attributes can be saved with the chart. For example, if you expand the Scale axes portion of the tree, you can see all of the attributes of data value labels that the template will include. You can select any attribute to include it in the template.

▶ Select All settings to include all of the available chart attributes in the template.

You can also enter a description of the template. This description will be visible when you apply the template.

Figure 6-15
Save Chart Template dialog box

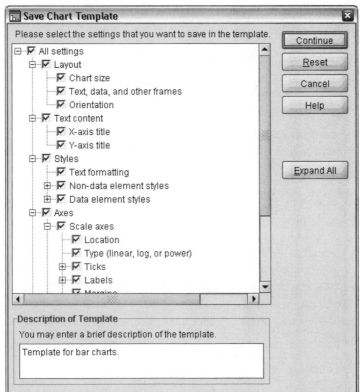

▶ Click Continue.

▶ In the Save Template dialog box, specify a location and filename for the template.

▶ When you are finished, click Save.

You can apply the template when you create a chart or in the Chart Editor. In the following example, we will apply it while creating a chart.

▶ Close the Chart Editor. The updated bar chart is shown in the Viewer.

Figure 6-16
Updated bar chart in Viewer

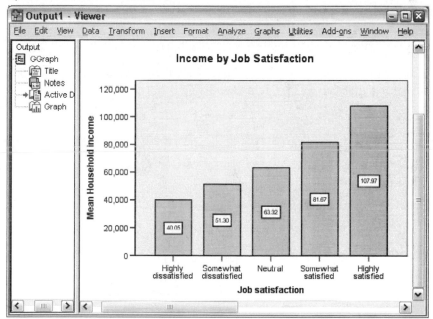

▶ From the Viewer menus choose:
Graphs
  Chart Builder...

The Chart Builder dialog box "remembers" the variables that you entered when you created the original chart. However, here you will create a slightly different chart to see how applying a template formats a chart.

▶ Remove *Job satisfaction* from the *x* axis by dragging it from the drop zone back to the Variables list. You can also click the drop zone and press Delete.

▶ Right-click *Level of education* in the Variables list and choose Ordinal.

▶ Drag *Level of education* from the Variables list to the *x* axis drop zone.

Because the title is now inaccurate, we are going to delete it.

▶ On the Titles/Footnotes tab, deselect Title 1.

Now we are going to specify the template to apply to the new chart.

▶ Click Options.

▶ In the Templates group in the Options dialog box, click Add.

▶ In the Find Template Files dialog box, locate the template file that you previously saved using the Save Chart Template dialog box.

▶ Select that file and click Open.

Figure 6-17
Options dialog box with template

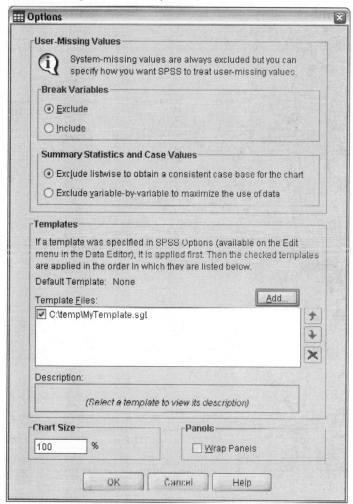

The Options dialog box displays the file path of the template you selected.

(Our example shows the path *C:\Program Files\SPSS\Looks\My Template.sgt*.)

▶ Click OK to close the Options dialog box.

Figure 6-18
Chart Builder with completed drop zones

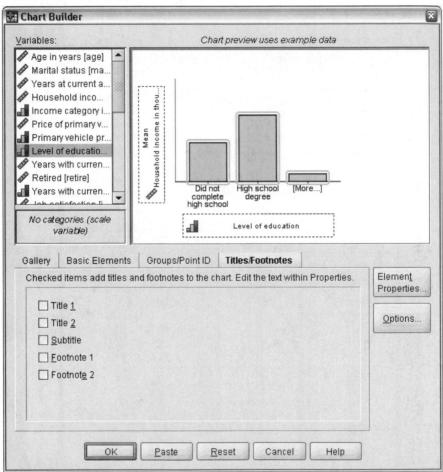

▶ Click OK on the Chart Builder dialog box to create the chart and apply the template.

The formatting in the new chart matches the formatting in the chart that you previously created and edited. Although the variables on the x axis are different, the charts otherwise resemble each other. Notice that the title from the previous chart was preserved in the template, even though you deleted the title in the Chart Builder.

If you want to apply templates after you've created a chart, you can do it in the Chart Editor (from the File menu, choose Apply Chart Template).

Figure 6-19
Updated bar chart in Viewer

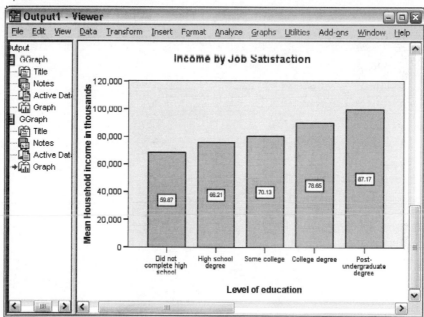

Defining Chart Options

In addition to using templates to format charts, you can use the Options to control various aspects of how charts are created.

▶ From the Data Editor or Viewer menus choose:
Edit
Options...

The Options dialog box contains many configuration settings. Click the Charts tab to see the available options.

Figure 6-20
Charts tab in Options dialog box

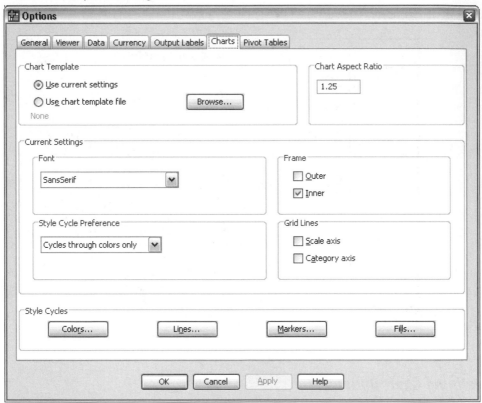

The options control how a chart is created. For each new chart, you can specify:

■ Whether to use the current settings or a template.

■ The width-to-height ratio (aspect ratio).

■ If you're not using a template, the settings to use for formatting.

■ The style cycles for graphic elements.

Style cycles allow you to specify the style of graphic elements in new charts. In this example, we'll look at the details for the color style cycle.

▶ Click Colors to open the Data Element Colors dialog box.

For a simple chart, the Chart Editor uses one style that you specify. For grouped charts, the Chart Editor uses a set of styles that it cycles through for each group (category) in the chart.

▶ Select Simple Charts.

▶ Select the light green color, which is third from the right in the second row from the bottom.

Figure 6-21
Data Element Colors dialog box

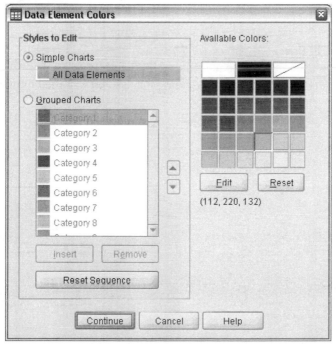

▶ Click Continue.

▶ In the Options dialog box, click OK to save the color style cycle changes.

The graphic elements in any new simple charts will now be light green.

▶ From the Data Editor or Viewer menus choose:
Graphs
  Chart Builder...

The Chart Builder displays the last chart you created. Remember that this chart had a template associated with it. We no longer want to use that template.

▶ Click Options.

▶ Deselect (uncheck) the template that you added previously. Note that you could also click the red *X* to delete the template. By deselecting rather than deleting, you keep the template available to use at another time.

▶ Click OK to create the chart.

The bars in the new chart are light green. This chart also differs from the last one in other ways. There is no title; the axis labels are in thousands; and there are no data labels. The differences occurred because the template wasn't applied to this chart.

Figure 6-22
Updated bar chart in Viewer

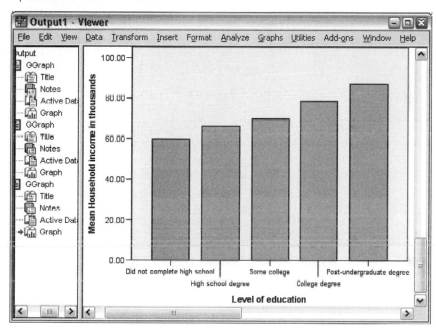

Working with Output

The results from running a statistical procedure are displayed in the Viewer. The output produced can be statistical tables, charts, graphs, or text, depending on the choices you make when you run the procedure. This section uses the files *viewertut.spv* and *demo.sav*. For more information, see "Sample Files" in Appendix A on p. 201.

Using the Viewer

Figure 7-1
Viewer

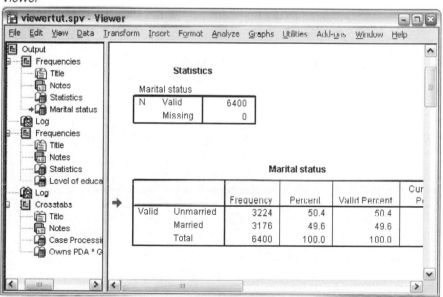

The Viewer window is divided into two panes. The **outline pane** contains an outline of all of the information stored in the Viewer. The **contents pane** contains statistical tables, charts, and text output.

Use the scroll bars to navigate through the window's contents, both vertically and horizontally. For easier navigation, click an item in the outline pane to display it in the contents pane.

If you find that there isn't enough room in the Viewer to see an entire table or that the outline view is too narrow, you can easily resize the window.

▶ Click and drag the right border of the outline pane to change its width.

An open book icon in the outline pane indicates that it is currently visible in the Viewer, although it may not currently be in the visible portion of the contents pane.

▶ To hide a table or chart, double-click its book icon in the outline pane.

The open book icon changes to a closed book icon, signifying that the information associated with it is now hidden.

▶ To redisplay the hidden output, double-click the closed book icon.

You can also hide all of the output from a particular statistical procedure or all of the output in the Viewer.

▶ Click the box with the minus sign (−) to the left of the procedure whose results you want to hide, or click the box next to the topmost item in the outline pane to hide all of the output.

The outline collapses, visually indicating that these results are hidden.

You can also change the order in which the output is displayed.

▶ In the outline pane, click on the items that you want to move.

▶ Drag the selected items to a new location in the outline and release the mouse button.

Figure 7-2
Reordered output in the Viewer

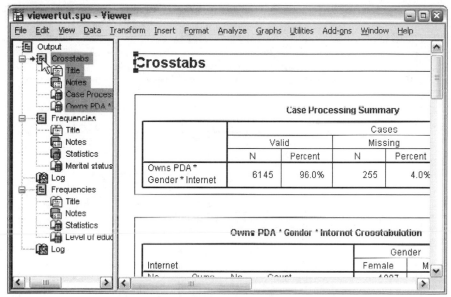

You can also move output items by clicking and dragging them in the contents pane.

Using the Pivot Table Editor

The results from most statistical procedures are displayed in **pivot tables**.

Accessing Output Definitions

Many statistical terms are displayed in the output. Definitions of these terms can be accessed directly in the Viewer.

▶ Double-click the *Owns PDA * Gender * Internet Crosstabulation* table.

▶ Right-click *Expected Count* and choose What's This? from the pop-up context menu.

The definition is displayed in a pop-up window.

Figure 7-3
Pop-up definition

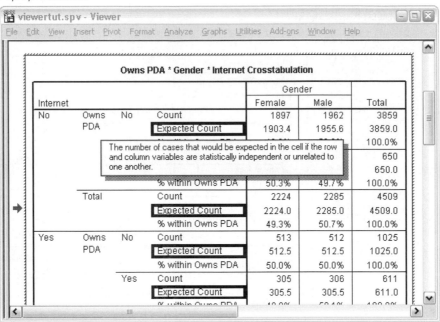

Pivoting Tables

The default tables produced may not display information as neatly or as clearly as you would like. With pivot tables, you can transpose rows and columns ("flip" the table), adjust the order of data in a table, and modify the table in many other ways. For example, you can change a short, wide table into a long, thin one by transposing rows and columns. Changing the layout of the table does not affect the results. Instead, it's a way to display your information in a different or more desirable manner.

▶ If it's not already activated, double-click the *Owns PDA * Gender * Internet Crosstabulation* table to activate it.

▶ If the Pivoting Trays window is not visible, from the menus choose:

Pivot
  Pivoting Trays

Pivoting trays provide a way to move data between columns, rows, and layers.

Figure 7-4
Pivoting trays

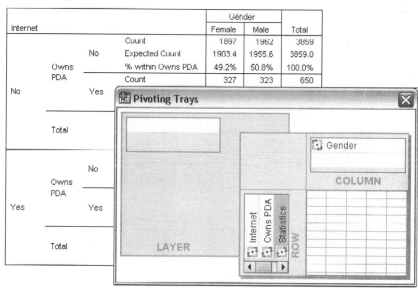

▶ Drag the *Statistics* element from the Row dimension to the Column dimension, below *Gender*. The table is immediately reconfigured to reflect your changes.

Figure 7-5
Moving rows to columns

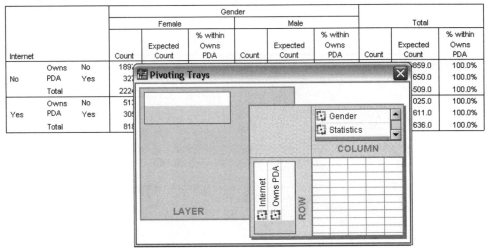

Internet					Gender						Total		
				Female				Male					
			Count	Expected Count	% within Owns PDA	Count	Expected Count	% within Owns PDA	Count	Expected Count	% within Owns PDA		
No	Owns PDA	No	189								859.0	100.0%	
		Yes	32								650.0	100.0%	
	Total		222								509.0	100.0%	
Yes	Owns PDA	No	51								025.0	100.0%	
		Yes	30								611.0	100.0%	
	Total		81								636.0	100.0%	

The order of the elements in the pivoting tray reflects the order of the elements in the table.

▶ Drag and drop the *Owns PDA* element before the *Internet* element in the row dimension to reverse the order of these two rows.

Figure 7-6
Swap rows

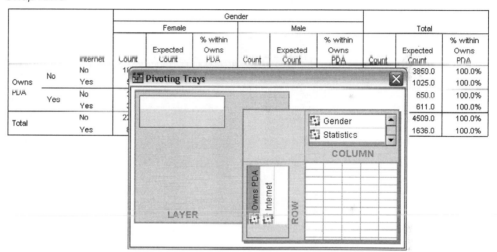

			Gender						Total		
			Female			Male					
		Internet	Count	Expected Count	% within Owns PDA	Count	Expected Count	% within Owns PDA	Count	Expected Count	% within Owns PDA
Owns PDA	No	No								3859.0	100.0%
		Yes								1025.0	100.0%
	Yes	No								650.0	100.0%
		Yes								611.0	100.0%
Total		No								4509.0	100.0%
		Yes								1636.0	100.0%

Creating and Displaying Layers

Layers can be useful for large tables with nested categories of information. By creating layers, you simplify the look of the table, making it easier to read.

▶ Drag the *Gender* element from the Column dimension to the Layer dimension.

Figure 7-7
Gender pivot icon in the Layer dimension

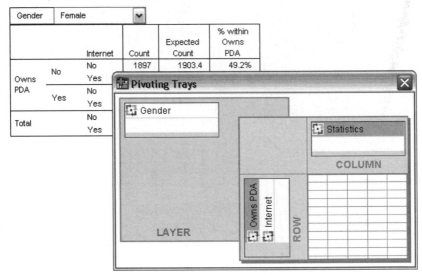

To display a different layer, select a category from the drop-down list in the table.

Figure 7-8
Choosing a layer

Gender	Gender Male ▼			
	Gender Female			% within
	Gender Male		Expected	Owns
	Total	t	Count	PDA
Owns PDA	No	No	1897	1903.4	49.2%
		Yes	513	512.5	50.0%
	Yes	No	327	320.6	50.3%
		Yes	305	305.5	49.9%
Total		No	2224	2224.0	49.3%
		Yes	818	818.0	50.0%

Editing Tables

Unless you've taken the time to create a custom TableLook, pivot tables are created with standard formatting. You can change the formatting of any text within a table. Formats that you can change include font name, font size, font style (bold or italic), and color.

▶ Double-click the *Level of education* table.

▶ If the Formatting toolbar is not visible, from the menus choose:
View
  Toolbar

▶ Click the title text, *Level of education*.

▶ From the drop-down list of font sizes on the toolbar, choose 12.

▶ To change the color of the title text, click the text color tool and choose a new color.

Figure 7-9
Reformatted title text in the pivot table

Level of education				
		Frequency	Percent	Valid Percent	Cumulative Percent
Valid	Did not complete high school	1390	21.7	21.7	21.7
	High school degree	1936	30.3	30.3	52.0
	Some college	1360	21.3	21.3	73.2
	College degree	1355	21.2	21.2	94.4
	Post-undergraduate degree	359	5.6	5.6	100.0
	Total	6400	100.0	100.0	

You can also edit the contents of tables and labels. For example, you can change the title of this table.

▶ Double-click the title.

▶ Type Education Level for the new label.

Note: If you change the values in a table, totals and other statistics are not recalculated.

Hiding Rows and Columns

Some of the data displayed in a table may not be useful or it may unnecessarily complicate the table. Fortunately, you can hide entire rows and columns without losing any data.

▶ If it's not already activated, double-click the *Education Level* table to activate it.

▶ Click *Valid Percent* column label to select it.

▶ From the Edit menu or the right-click context menu choose:
Select
  Data and Label Cells

▶ From the View menu choose Hide or from the right-click context menu choose Hide Category.

The column is now hidden but not deleted.

Figure 7-10
Valid Percent column hidden in table

Education Level

		Frequency	Percent	Cumulative Percent
Valid	Did not complete high school	1390	21.7	21.7
	High school degree	1936	30.3	52.0
	Some college	1360	21.3	73.2
	College degree	1355	21.2	94.4
	Post-undergraduate degree	359	5.6	100.0
	Total	6400	100.0	

To redisplay the column:

▶ From the menus choose:
View
  Show All

Rows can be hidden and displayed in the same way as columns.

Changing Data Display Formats

You can easily change the display format of data in pivot tables.

▶ If it's not already activated, double-click the *Education Level* table to activate it.

▶ Click on the *Percent* column label to select it.

▶ From the Edit menu or the right-click context menu choose:
Select
  Data Cells

▶ From the Format menu or the right-click context menu choose Cell Properties.

▶ Click the Format Value tab.

Chapter 7

▶ Type 0 in the Decimals field to hide all decimal points in this column.

Figure 7-11
Cell Properties, Format Value tab

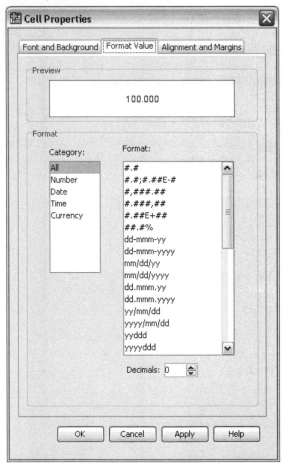

You can also change the data type and format in this dialog box.

▶ Select the type that you want from the Category list, and then select the format for that type in the Format list.

▶ Click OK or Apply to apply your changes.

Figure 7-12
Decimals hidden in Percent column

Education Level

		Frequency	Percent	Cumulative Percent
Valid	Did not complete high school	1390	22	21.7
	High school degree	1936	30	52.0
	Some college	1360	21	73.2
	College degree	1355	21	94.4
	Post-undergraduate degree	359	6	100.0
	Total	6400	100	

The decimals are now hidden in the *Percent* column.

TableLooks

The format of your tables is a critical part of providing clear, concise, and meaningful results. If your table is difficult to read, the information contained within that table may not be easily understood.

Using Predefined Formats

▶ Double-click the *Marital status* table.

▶ From the menus choose:
Format
  TableLooks...

The TableLooks dialog box lists a variety of predefined styles. Select a style from the list to preview it in the Sample window on the right.

Figure 7-13
TableLooks dialog box

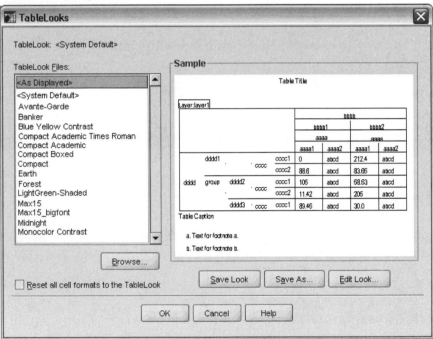

You can use a style as is, or you can edit an existing style to better suit your needs.

▶ To use an existing style, select one and click OK.

Customizing TableLook Styles

You can customize a format to fit your specific needs. Almost all aspects of a table can be customized, from the background color to the border styles.

▶ Double-click the *Marital status* table.

▶ From the menus choose:

Format
  TableLooks...

▶ Select the style that is closest to your desired format and click Edit Look.

▶ Click the Cell Formats tab to view the formatting options.

Figure 7-14
Table Properties dialog box

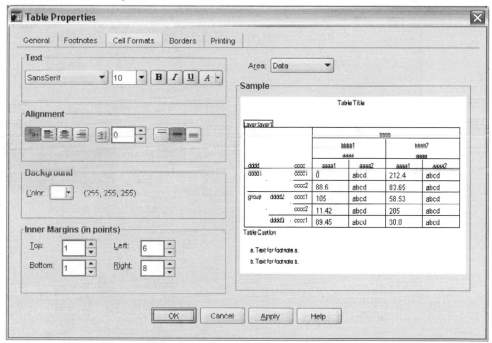

The formatting options include font name, font size, style, and color. Additional options include alignment, text and background colors, and margin sizes.

The Sample window on the right provides a preview of how the formatting changes affect your table. Each area of the table can have different formatting styles. For example, you probably wouldn't want the title to have the same style as the data. To select a table area to edit, you can either choose the area by name in the Area drop-down list, or you can click the area that you want to change in the Sample window.

▶ Select Data from the Area drop-down list.

▶ Select a new color from the Background drop-down palette.

▶ Then select a new text color.

The Sample window shows the new style.

Figure 7-15
Changing table cell formats

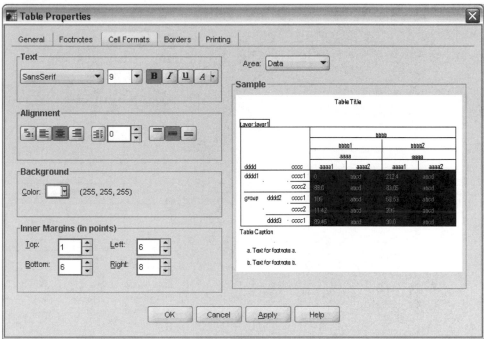

▶ Click OK to return to the TableLooks dialog box.

You can save your new style, which allows you to apply it to future tables easily.

▶ Click Save As.

▶ Navigate to the desired target directory and enter a name for your new style in the File Name text box.

▶ Click Save.

▶ Click OK to apply your changes and return to the Viewer.

The table now contains the custom formatting that you specified.

Figure 7-16
Custom TableLook

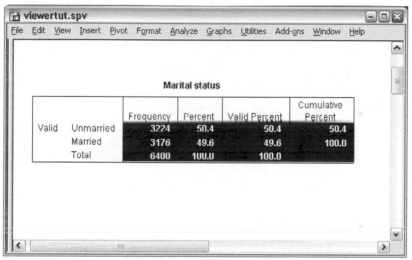

Changing the Default Table Formats

Although you can change the format of a table after it has been created, it may be more efficient to change the default TableLook so that you do not have to change the format every time you create a table.

To change the default TableLook style for your pivot tables, from the menus choose:
Edit
  Options...

▶ Click the Pivot Tables tab in the Options dialog box.

Figure 7-17
Options dialog box

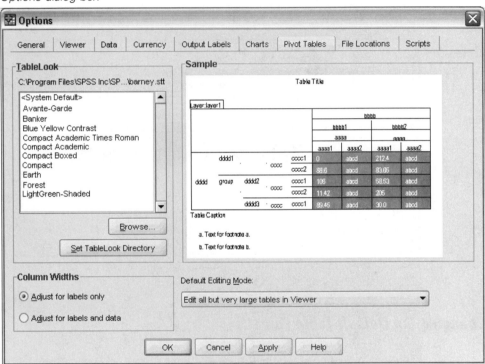

▶ Select the TableLook style that you want to use for all new tables.

The Sample window on the right shows a preview of each TableLook.

▶ Click OK to save your settings and close the dialog box.

All tables that you create after changing the default TableLook automatically conform to the new formatting rules.

Customizing the Initial Display Settings

The initial display settings include the alignment of objects in the Viewer, whether objects are shown or hidden by default, and the width of the Viewer window. To change these settings:

▶ From the menus choose:
Edit
  Options...

▶ Click the Viewer tab.

Figure 7-18
Viewer options

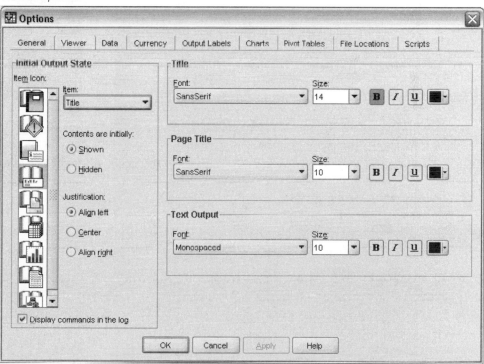

The settings are applied on an object-by-object basis. For example, you can customize the way charts are displayed without making any changes to the way tables are displayed. Simply select the object that you want to customize, and make the desired changes.

▶ Click the Title icon to display its settings.

▶ Click Center to display all titles in the (horizontal) center of the Viewer.

You can also hide elements, such as the log and warning messages, that tend to clutter your output. Double-clicking on an icon automatically changes that object's display property.

▶ Double-click the Warnings icon to hide warning messages in the output.

▶ Click OK to save your changes and close the dialog box.

Displaying Variable and Value Labels

In most cases, displaying the labels for variables and values is more effective than displaying the variable name or the actual data value. There may be cases, however, when you want to display both the names and the labels.

▶ From the menus choose:
  Edit
    Options...

▶ Click the Output Labels tab.

Figure 7-19
Output Labels options

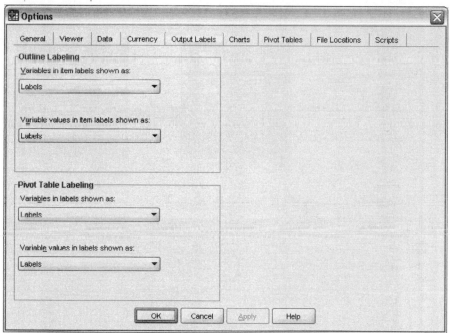

You can specify different settings for the outline and contents panes. For example, to show labels in the outline and variable names and data values in the contents:

▶ In the Pivot Table Labeling group, select Names from the Variables in Labels drop-down list to show variable names instead of labels.

▶ Then, select Values from the Variable Values in Labels drop-down list to show data values instead of labels.

Figure 7-20
Pivot Table Labeling settings

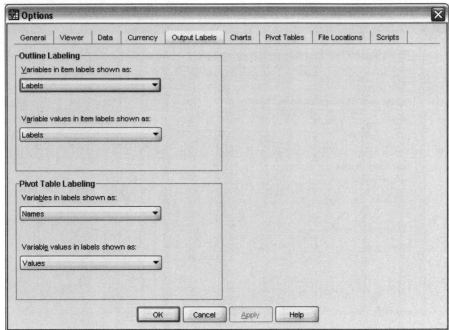

Subsequent tables produced in the session will reflect these changes.

Figure 7-21
Variable names and values displayed

marital

		Frequency	Percent	Valid Percent	Cumulative Percent
Valid	0	3224	50.4	50.4	50.4
	1	3176	49.6	49.6	100.0
	Total	6400	100.0	100.0	

Using Results in Other Applications

Your results can be used in many applications. For example, you may want to include a table or chart in a presentation or report.

The following examples are specific to Microsoft Word, but they may work similarly in other word processing applications.

Pasting Results as Word Tables

You can paste pivot tables into Word as native Word tables. All table attributes, such as font sizes and colors, are retained. Because the table is pasted in the Word table format, you can edit it in Word just like any other table.

▶ Click the *Marital status* table in the Viewer.

▶ From the menus choose:
Edit
  Copy

▶ Open your word processing application.

▶ From the word processor's menus choose:
Edit
  Paste Special...

▶ Select Formatted Text (RTF) in the Paste Special dialog box.

Figure 7-22
Paste Special dialog box

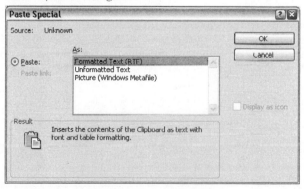

▶ Click OK to paste your results into the current document.

Figure 7-23
Pivot table displayed in Word

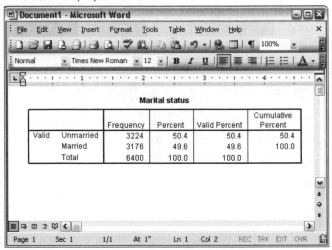

The table is now displayed in your document. You can apply custom formatting, edit the data, and resize the table to fit your needs.

Pasting Results as Text

Pivot tables can be copied to other applications as plain text. Formatting styles are not retained in this method, but you can edit the table data after you paste it into the target application.

▶ Click the *Marital status* table in the Viewer.

▶ From the menus choose:
Edit
  Copy

▶ Open your word processing application.

▶ From the word processor's menus choose:
Edit
  Paste Special...

▶ Select Unformatted Text in the Paste Special dialog box.

Figure 7-24
Paste Special dialog box

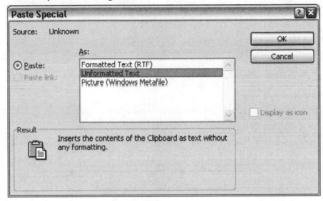

▶ Click OK to paste your results into the current document.

Figure 7-25
Pivot table displayed in Word

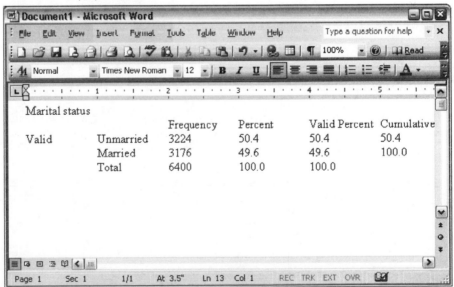

Each column of the table is separated by tabs. You can change the column widths by adjusting the tab stops in your word processing application.

Exporting Results to Microsoft Word, PowerPoint, and Excel Files

You can export results to a Microsoft Word , PowerPoint, or Excel file. You can export selected items or all items in the Viewer. This section uses the files *msouttut.spv* and *demo.sav*. For more information, see "Sample Files" in Appendix A on p. 201.

Note: Export to PowerPoint is available only on Windows operating systems and is not available with the Student Version.

In the Viewer's outline pane, you can select specific items that you want to export. You do not have to select specific items.

▶ From the Viewer menus choose:
File
  Export...

Instead of exporting all objects in the Viewer, you can choose to export only visible objects (open books in the outline pane) or those that you selected in the outline pane. If you did not select any items in the outline pane, you do not have the option to export selected objects.

Figure 7-26
Export Output dialog box

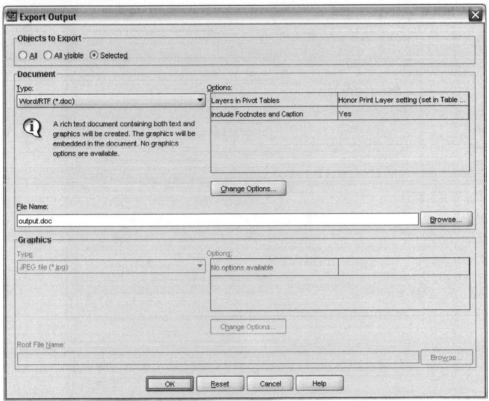

▶ In the Objects to Export group, select All.

▶ From the Type drop-down list select Word/RTF file (*.doc).

▶ Click OK to generate the Word file.

When you open the resulting file in Word, you can see how the results are exported. Notes, which are not visible objects, appear in Word because you chose to export all objects.

Figure 7-27
Output.doc in Word

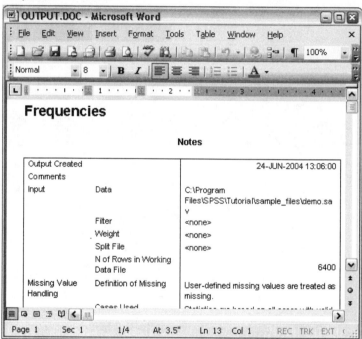

Pivot tables become Word tables, with all of the formatting of the original pivot table retained, including fonts, colors, borders, and so on.

Figure 7-28
Pivot tables in Word

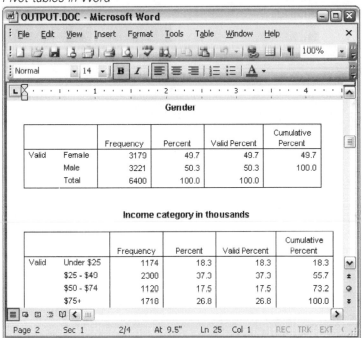

Charts are included in the Word document as graphic images.

Figure 7-29
Charts in Word

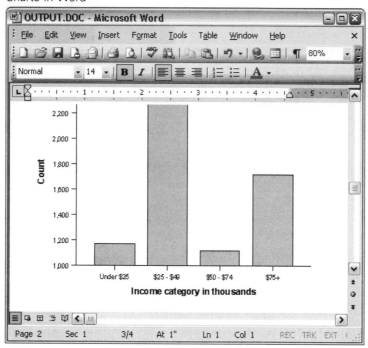

Text output is displayed in the same font used for the text object in the Viewer. For proper alignment, text output should use a fixed-pitch (monospaced) font.

Figure 7-30
Text output in Word

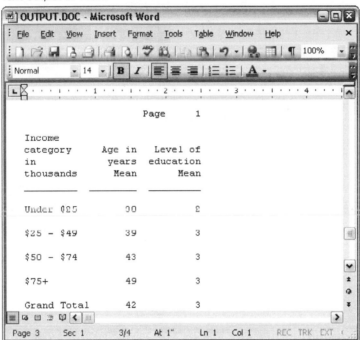

If you export to a PowerPoint file, each exported item is placed on a separate slide. Pivot tables exported to PowerPoint become Word tables, with all of the formatting of the original pivot table, including fonts, colors, borders, and so on.

Figure 7-31
Pivot tables in PowerPoint

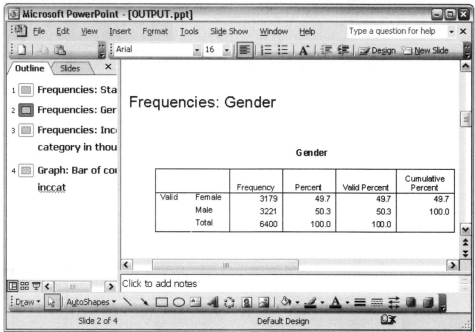

Charts selected for export to PowerPoint are embedded in the PowerPoint file.

Figure 7-32
Charts in PowerPoint

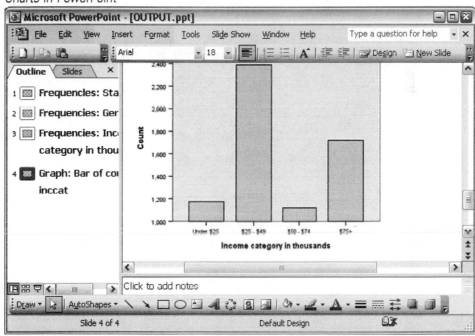

If you export to an Excel file, results are exported differently.

Figure 7-33
Output.xls in Excel

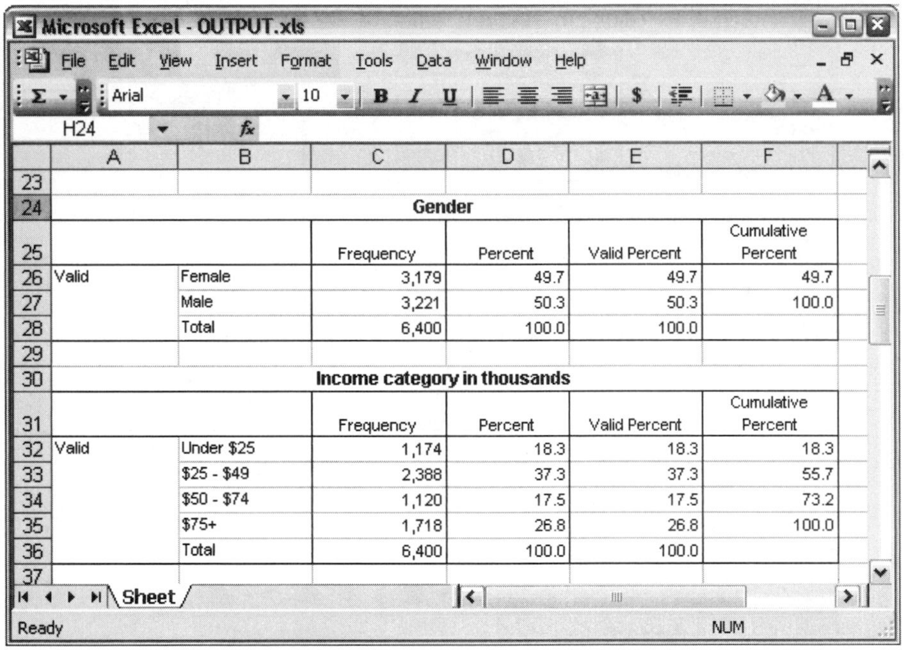

	A	B	C	D	E	F
23						
24	**Gender**					
25			Frequency	Percent	Valid Percent	Cumulative Percent
26	Valid	Female	3,179	49.7	49.7	49.7
27		Male	3,221	50.3	50.3	100.0
28		Total	6,400	100.0	100.0	
29						
30	**Income category in thousands**					
31			Frequency	Percent	Valid Percent	Cumulative Percent
32	Valid	Under $25	1,174	18.3	18.3	18.3
33		$25 - $49	2,388	37.3	37.3	55.7
34		$50 - $74	1,120	17.5	17.5	73.2
35		$75+	1,718	26.8	26.8	100.0
36		Total	6,400	100.0	100.0	
37						

Pivot table rows, columns, and cells become Excel rows, columns, and cells.

Each line in the text output is a row in the Excel file, with the entire contents of the line contained in a single cell. Charts are not exported at all.

Figure 7-34
Text output in Excel

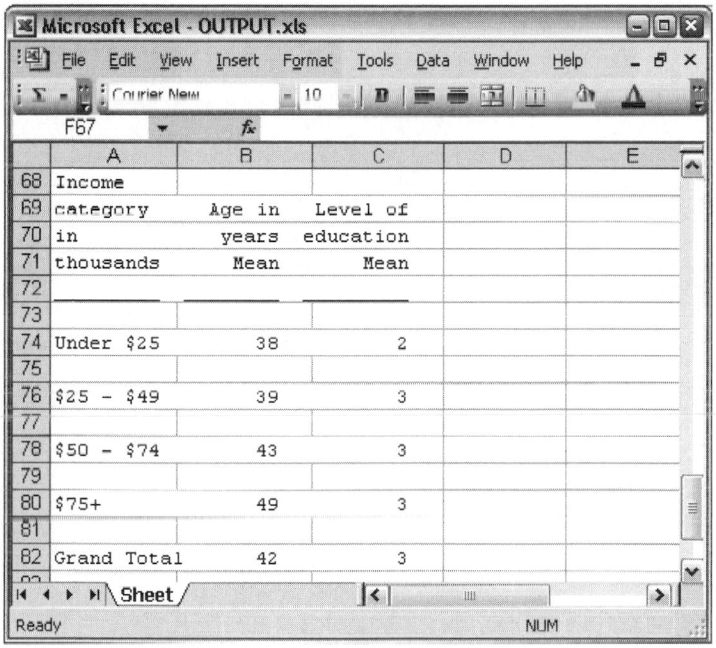

Exporting Results to PDF

You can export all or selected items in the Viewer to a PDF (portable document format) file.

▶ From the menus in the Viewer window that contains the result you want to export to PDF choose:

File
  Export...

▶ In the Export Output dialog box, from the Export Format File Type drop-down list choose Portable Document Format.

Figure 7-35
Export Output dialog box

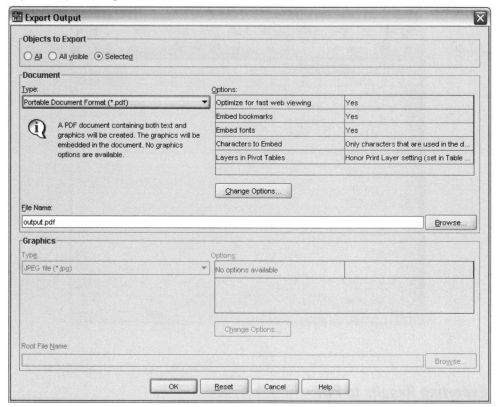

- The outline pane of the Viewer document is converted to bookmarks in the PDF file for easy navigation.

- Page size, orientation, margins, content and display of page headers and footers, and printed chart size in PDF documents are controlled by page setup options (File menu, Page Setup in the Viewer window).

- The resolution (DPI) of the PDF document is the current resolution setting for the default or currently selected printer (which can be changed using Page Setup). The maximum resolution is 1200 DPI. If the printer setting is higher, the PDF document resolution will be 1200 DPI. *Note*: High-resolution documents may yield poor results when printed on lower-resolution printers.

Figure 7-36
PDF file with bookmarks

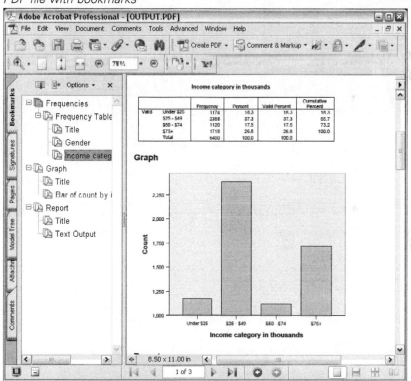

Exporting Results to HTML

You can also export results to HTML (hypertext markup language). When saving as HTML, all non-graphic output is exported into a single HTML file.

Figure 7-37
Output.htm in Web browser

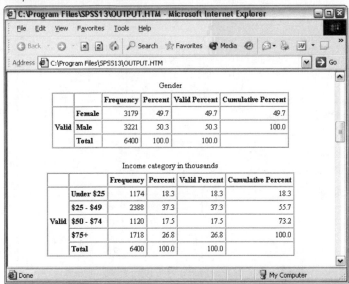

When you export to HTML, charts can be exported as well, but not to a single file.

Each chart will be saved as a file in a format that you specify, and references to these graphics files will be placed in the HTML. There is also an option to export all charts (or selected charts) in separate graphics files.

Figure 7-38
References to graphics in HTML output

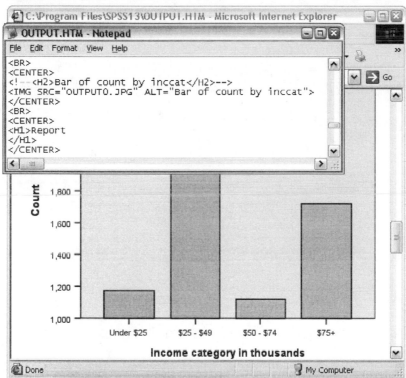

Working with Syntax

You can save and automate many common tasks by using the powerful command language. It also provides some functionality not found in the menus and dialog boxes. Most commands are accessible from the menus and dialog boxes. However, some commands and options are available only by using the command language. The command language also allows you to save your jobs in a syntax file so that you can repeat your analysis at a later date.

A command syntax file is simply a text file that contains SPSS syntax commands. You can open a syntax window and type commands directly, but it is often easier to let the dialog boxes do some or all of the work for you.

The examples in this chapter use the data file *demo.sav*. For more information, see "Sample Files" in Appendix Λ on p. 201.

Note: Command syntax is not available with the Student Version.

Pasting Syntax

The easiest way to create syntax is to use the Paste button located on most dialog boxes.

▶ Open *demo.sav* for use in this example.

▶ From the menus choose:
Analyze
  Descriptive Statistics
    Frequencies...

The Frequencies dialog box opens.

Figure 8-1
Frequencies dialog box

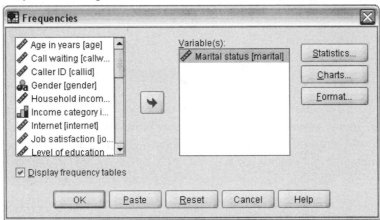

▶ Select *Marital status [marital]* and move it into the Variable(s) list.

▶ Click Charts.

▶ In the Charts dialog box, select Bar charts.

▶ In the Chart Values group, select Percentages.

▶ Click Continue.

▶ Click Paste to copy the syntax created as a result of the dialog box selections to the Syntax Editor.

Figure 8-2
Frequencies syntax

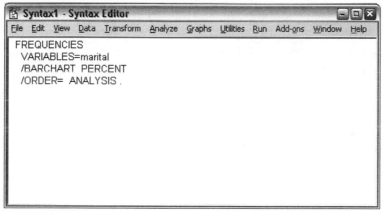

▶ To run the syntax currently displayed, from the menus choose:

Run
  Current

Editing Syntax

In the syntax window, you can edit the syntax. For example, you could change the subcommand /BARCHART to display frequencies instead of percentages. (A subcommand is indicated by a slash.)

Figure 8-3
Modified syntax

To find out what subcommands and keywords are available for the current command, press the F1 key. This takes you directly to the command syntax reference information for the current command.

Figure 8-4
FREQUENCIES syntax help

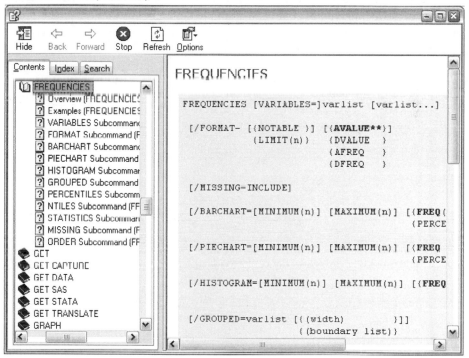

Opening and Running a Syntax File

▶ To open a saved syntax file, from the menus choose:

File
  Open
    Syntax...

A standard dialog box for opening files is displayed.

▶ Select a syntax file. If no syntax files are displayed, make sure Syntax (*.sps) is selected as the file type you want to view.

▶ Click Open.

▶ Use the Run menu in the syntax window to run the commands.

If the commands apply to a specific data file, the data file must be opened before running the commands, or you must include a command that opens the data file. You can paste this type of command from the dialog boxes that open data files.

Modifying Data Values

The data you start with may not always be organized in the most useful manner for your analysis or reporting needs. For example, you may want to:

- Create a categorical variable from a scale variable.

- Combine several response categories into a single category.

- Create a new variable that is the computed difference between two existing variables.

- Calculate the length of time between two dates.

This chapter uses the data file *demo.sav*. For more information, see "Sample Files" in Appendix A on p. 201.

Creating a Categorical Variable from a Scale Variable

Several categorical variables in the data file *demo.sav* are, in fact, derived from scale variables in that data file. For example, the variable *inccat* is simply *income* grouped into four categories. This categorical variable uses the integer values 1–4 to represent the following income categories (in thousands): less than $25, $25–$49, $50–$74, and $75 or higher.

To create the categorical variable *inccat*:

▶ From the menus in the Data Editor window choose:
Transform
　Visual Binning...

Figure 9-1
Initial Visual Binning dialog box

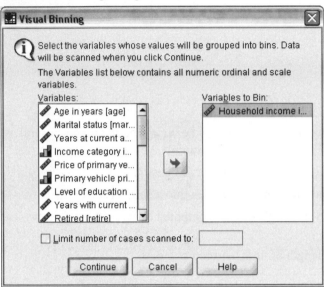

In the initial Visual Binning dialog box, you select the scale and/or ordinal variables for which you want to create new, binned variables. **Binning** means taking two or more contiguous values and grouping them into the same category.

Since Visual Binning relies on actual values in the data file to help you make good binning choices, it needs to read the data file first. Since this can take some time if your data file contains a large number of cases, this initial dialog box also allows you to limit the number of cases to read ("scan"). This is not necessary for our sample data file. Even though it contains more than 6,000 cases, it does not take long to scan that number of cases.

▶ Drag and drop *Household income in thousands [income]* from the Variables list into the Variables to Bin list, and then click Continue.

Figure 9-2
Main Visual Binning dialog box

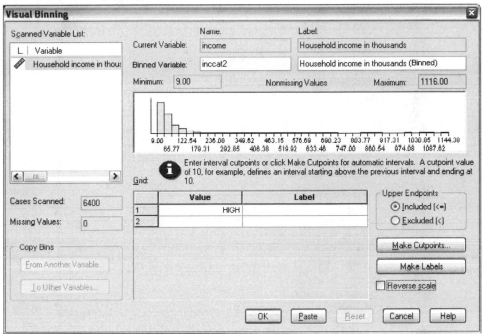

▶ In the main Visual Binning dialog box, select *Household income in thousands [income]* in the Scanned Variable List.

A histogram displays the distribution of the selected variable (which in this case is highly skewed).

▶ Enter inccat2 for the new binned variable name and Income category [in thousands] for the variable label.

▶ Click Make Cutpoints.

Figure 9-3
Visual Binning Cutpoints dialog box

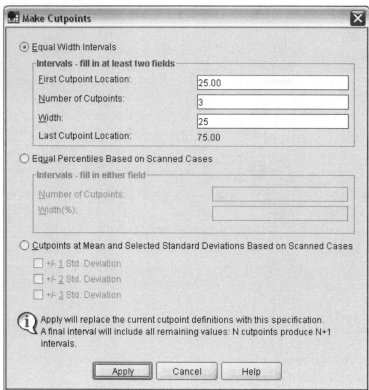

▶ Select Equal Width Intervals.

▶ Enter 25 for the first cutpoint location, 3 for the number of cutpoints, and 25 for the width.

The number of binned categories is one greater than the number of cutpoints. So in this example, the new binned variable will have four categories, with the first three categories each containing ranges of 25 (thousand) and the last one containing all values above the highest cutpoint value of 75 (thousand).

▶ Click Apply.

Figure 9-4
Main Visual Binning dialog box with defined cutpoints

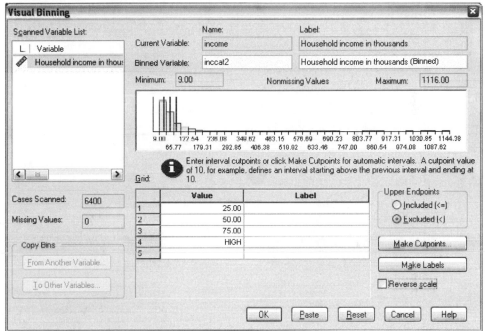

The values now displayed in the grid represent the defined cutpoints, which are the upper endpoints of each category. Vertical lines in the histogram also indicate the locations of the cutpoints.

By default, these cutpoint values are included in the corresponding categories. For example, the first value of 25 would include all values less than or equal to 25. But in this example, we want categories that correspond to less than 25, 25–49, 50–74, and 75 or higher.

▶ In the Upper Endpoints group, select Excluded (<).

▶ Then click Make Labels.

Figure 9-5
Automatically generated value labels

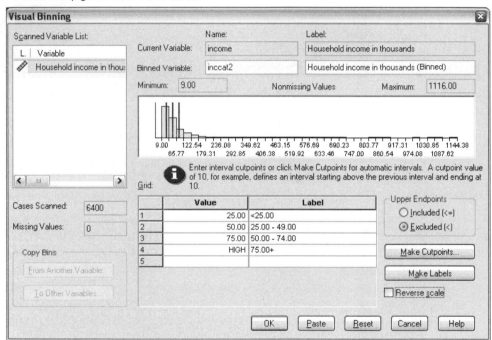

This automatically generates descriptive value labels for each category. Since the actual values assigned to the new binned variable are simply sequential integers starting with 1, the value labels can be very useful.

You can also manually enter or change cutpoints and labels in the grid, change cutpoint locations by dragging and dropping the cutpoint lines in the histogram, and delete cutpoints by dragging cutpoint lines off of the histogram.

► Click OK to create the new, binned variable.

The new variable is displayed in the Data Editor. Since the variable is added to the end of the file, it is displayed in the far right column in Data View and in the last row in Variable View.

Figure 9-6
New variable displayed in Data Editor

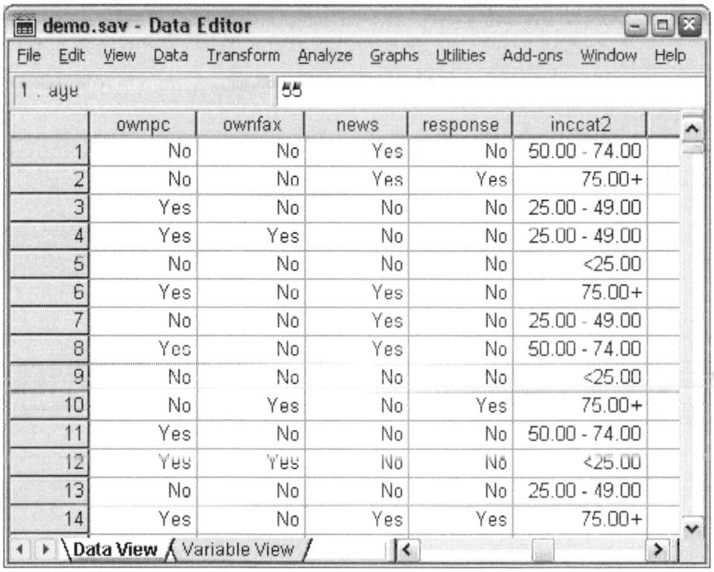
demo.sav - Data Editor
File Edit View Data Transform Analyze Graphs Utilities Add-ons Window Help
1 : age
55
ownpc ownfax news response inccat2
1 No No Yes No 50.00 - 74.00
2 No No Yes Yes 75.00+
3 Yes No No No 25.00 - 49.00
4 Yes Yes No No 25.00 - 49.00
5 No No No No <25.00
6 Yes No Yes No 75.00+
7 No No Yes No 25.00 - 49.00
8 Yes No Yes No 50.00 - 74.00
9 No No No No <25.00
10 No Yes No Yes 75.00+
11 Yes No No No 50.00 - 74.00
12 Yes Yes No No <25.00
13 No No No No 25.00 - 49.00
14 Yes No Yes Yes 75.00+
Data View
Variable View

Computing New Variables

Using a wide variety of mathematical functions, you can compute new variables based on highly complex equations. In this example, however, we will simply compute a new variable that is the difference between the values of two existing variables.

The data file *demo.sav* contains a variable for the respondent's current age and a variable for the number of years at current job. It does not, however, contain a variable for the respondent's age at the time he or she started that job. We can create a new variable that is the computed difference between current age and number of years at current job, which should be the approximate age at which the respondent started that job.

▶ From the menus in the Data Editor window choose:

Transform
  Compute Variable...

▶ For Target Variable, enter jobstart.

▶ Select *Age in years [age]* in the source variable list and click the arrow button to copy it to the Numeric Expression text box.

▶ Click the minus (–) button on the calculator pad in the dialog box (or press the minus key on the keyboard).

▶ Select *Years with current employer [employ]* and click the arrow button to copy it to the expression.

Figure 9-7
Compute Variable dialog box

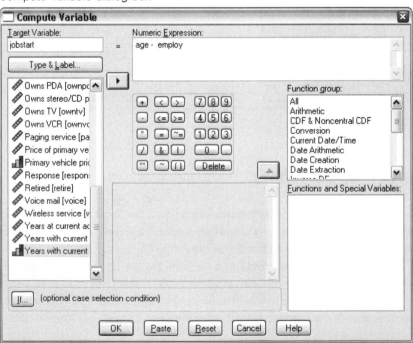

Note: Be careful to select the correct employment variable. There is also a recoded categorical version of the variable, which is *not* what you want. The numeric expression should be *age–employ*, not *age–empcat*.

▶ Click OK to compute the new variable.

The new variable is displayed in the Data Editor. Since the variable is added to the end of the file, it is displayed in the far right column in Data View and in the last row in Variable View.

Figure 9-8
New variable displayed in Data Editor

	ownfax	news	response	inccat2	jobstart	
1	No	Yes	No	50.00 - 74.00	52.00	
2	No	Yes	Yes	75.00+	53.00	
3	No	No	No	25.00 - 49.00	27.00	
4	Yes	No	No	25.00 - 49.00	23.00	
5	No	No	No	<25.00	23.00	
6	No	Yes	No	75.00+	43.00	
7	No	Yes	No	25.00 - 49.00	40.00	
8	No	Yes	No	50.00 - 74.00	34.00	
9	No	No	No	<25.00	44.00	
10	Yes	No	Yes	75.00+	32.00	
11	No	No	No	50.00 - 74.00	54.00	
12	Yes	No	No	<25.00	27.00	
13	No	No	No	25.00 - 49.00	30.00	
14	No	Yes	Yes	75.00+	41.00	

demo.sav - Data Editor

File   Edit   View   Data   Transform   Analyze   Graphs   Utilities   Add-ons   Window   Help

1 : age   55

Data View Variable View

Using Functions in Expressions

You can also use predefined functions in expressions. More than 70 built-in functions are available, including:

- Arithmetic functions
- Statistical functions
- Distribution functions
- Logical functions

- Date and time aggregation and extraction functions
- Missing-value functions
- Cross-case functions
- String functions

Figure 9-9
Compute Variable dialog box displaying function grouping

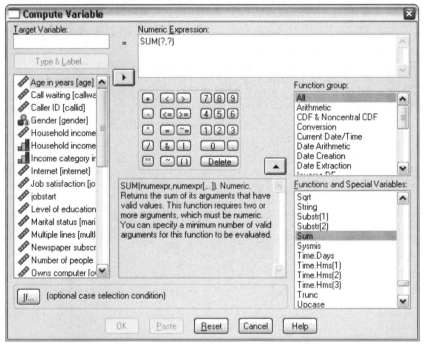

Functions are organized into logically distinct groups, such as a group for arithmetic operations and another for computing statistical metrics. For convenience, a number of commonly used system variables, such as *$TIME* (current date and time), are also included in appropriate function groups. A brief description of the currently selected function (in this case, SUM) or system variable is displayed in a reserved area in the Compute Variable dialog box.

Pasting a Function into an Expression

To paste a function into an expression:

▶ Position the cursor in the expression at the point where you want the function to appear.

▶ Select the appropriate group from the Function group list. The group labeled All provides a listing of all available functions and system variables.

▶ Double-click the function in the Functions and Special Variables list (or select the function and click the arrow adjacent to the Function group list).

The function is inserted into the expression. If you highlight part of the expression and then insert the function, the highlighted portion of the expression is used as the first argument in the function.

Editing a Function in an Expression

The function is not complete until you enter the arguments, represented by question marks in the pasted function. The number of question marks indicates the minimum number of arguments required to complete the function.

▶ Highlight the question mark(s) in the pasted function.

▶ Enter the arguments. If the arguments are variable names, you can paste them from the variable list.

Using Conditional Expressions

You can use conditional expressions (also called logical expressions) to apply transformations to selected subsets of cases. A conditional expression returns a value of true, false, or missing for each case. If the result of a conditional expression is true, the transformation is applied to that case. If the result is false or missing, the transformation is not applied to the case.

To specify a conditional expression:

▶ Click If in the Compute Variable dialog box. This opens the If Cases dialog box.

Chapter 9

Figure 9-10
If Cases dialog box

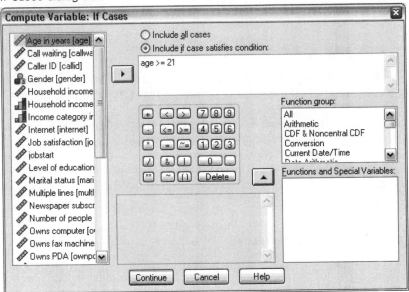

▶ Select Include if case satisfies condition.

▶ Enter the conditional expression.

Most conditional expressions contain at least one relational operator, as in:

```
age>=21
```

or

```
income*3<100
```

In the first example, only cases with a value of 21 or greater for *Age [age]* are selected. In the second example, *Household income in thousands [income]* multiplied by 3 must be less than 100 for a case to be selected.

You can also link two or more conditional expressions using logical operators, as in:

```
age>=21 | ed>=4
```

or

```
income*3<100 & ed=5
```

In the first example, cases that meet either the *Age [age]* condition or the *Level of education [ed]* condition are selected. In the second example, both the *Household income in thousands [income]* and *Level of education [ed]* conditions must be met for a case to be selected.

Working with Dates and Times

A number of tasks commonly performed with dates and times can be easily accomplished using the Date and Time Wizard. Using this wizard, you can:

■ Create a date/time variable from a string variable containing a date or time.

■ Construct a date/time variable by merging variables containing different parts of the date or time.

■ Add or subtract values from date/time variables, including adding or subtracting two date/time variables.

■ Extract a part of a date or time variable; for example, the day of month from a date/time variable which has the form mm/dd/yyyy.

The examples in this section use the data file *upgrade.sav*. For more information, see "Sample Files" in Appendix A on p. 201.

To use the Date and Time Wizard:

▶ From the menus choose:
Transform
  Date and Time Wizard...

Figure 9-11
Date and Time Wizard introduction screen

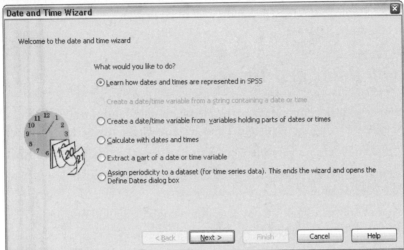

The introduction screen of the Date and Time Wizard presents you with a set of general tasks. Tasks that do not apply to the current data are disabled. For example, the data file *upgrade.sav* doesn't contain any string variables, so the task to create a date variable from a string is disabled.

If you're new to dates and times in SPSS, you can select Learn how dates and times are represented in SPSS and click Next. This leads to a screen that provides a brief overview of date/time variables and a link, through the Help button, to more detailed information.

Calculating the Length of Time between Two Dates

One of the most common tasks involving dates is calculating the length of time between two dates. As an example, consider a software company interested in analyzing purchases of upgrade licenses by determining the number of years since each customer last purchased an upgrade. The data file *upgrade.sav* contains a variable for the date on which each customer last purchased an upgrade but not the number of years since that purchase. A new variable that is the length of time in years between the date of the last upgrade and the date of the next product release will provide a measure of this quantity.

To calculate the length of time between two dates:

▶ Select Calculate with dates and times on the introduction screen of the Date and Time Wizard and click Next.

Figure 9-12
Calculating the length of time between two dates: Step 1

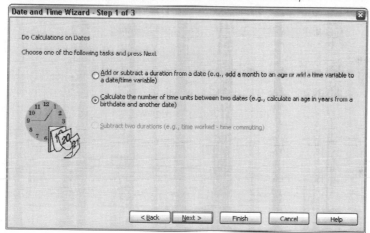

▶ Select Calculate the number of time units between two dates and click Next.

Figure 9-13
Calculating the length of time between two dates: Step 2

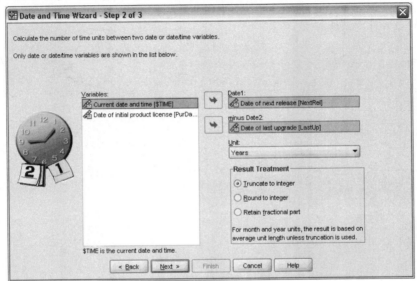

▶ Select *Date of next release* for Date1.

▶ Select *Date of last upgrade* for Date2.

▶ Select Years for the Unit and Truncate to Integer for the Result Treatment. (These are the default selections.)

▶ Click Next.

Figure 9-14
Calculating the length of time between two dates: Step 3

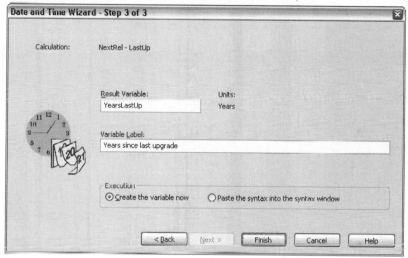

▶ Enter *YearsLastUp* for the name of the result variable. Result variables cannot have the same name as an existing variable.

▶ Enter *Years since last upgrade* as the label for the result variable. Variable labels for result variables are optional.

▶ Leave the default selection of Create the variable now, and click Finish to create the new variable.

The new variable, *YearsLastUp*, displayed in the Data Editor is the integer number of years between the two dates. Fractional parts of a year have been truncated.

Figure 9-15
New variable displayed in Data Editor

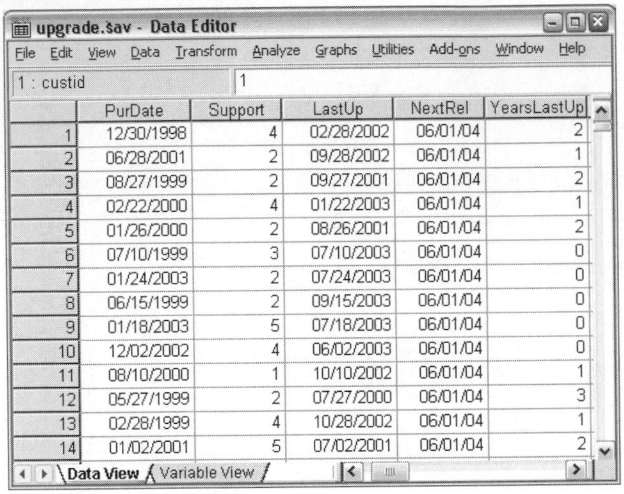

	PurDate	Support	LastUp	NextRel	YearsLastUp
1	12/30/1998	4	02/28/2002	06/01/04	2
2	06/28/2001	2	09/28/2002	06/01/04	1
3	08/27/1999	2	09/27/2001	06/01/04	2
4	02/22/2000	4	01/22/2003	06/01/04	1
5	01/26/2000	2	08/26/2001	06/01/04	2
6	07/10/1999	3	07/10/2003	06/01/04	0
7	01/24/2003	2	07/24/2003	06/01/04	0
8	06/15/1999	2	09/15/2003	06/01/04	0
9	01/18/2003	5	07/18/2003	06/01/04	0
10	12/02/2002	4	06/02/2003	06/01/04	0
11	08/10/2000	1	10/10/2002	06/01/04	1
12	05/27/1999	2	07/27/2000	06/01/04	3
13	02/28/1999	4	10/28/2002	06/01/04	1
14	01/02/2001	5	07/02/2001	06/01/04	2

Adding a Duration to a Date

You can add or subtract durations, such as 10 days or 12 months, to a date. Continuing with the example of the software company from the previous section, consider determining the date on which each customer's initial tech support contract ends. The data file *upgrade.sav* contains a variable for the number of years of contracted support and a variable for the initial purchase date. You can then determine the end date of the initial support by adding years of support to the purchase date.

To add a duration to a date:

▶ Select Calculate with dates and times on the introduction screen of the Date and Time Wizard and click Next.

▶ Select Add or subtract a duration from a date and click Next.

Figure 9-16
Adding a duration to a date: Step 2

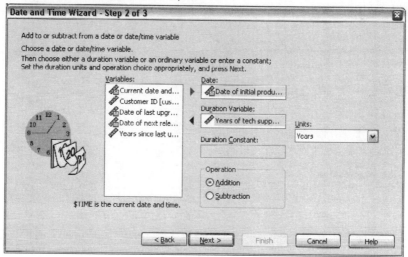

▶ Select *Date of initial product license* for Date.

▶ Select *Years of tech support* for the Duration Variable.

Since *Years of tech support* is simply a numeric variable, you need to indicate the units to use when adding this variable as a duration.

▶ Select Years from the Units drop-down list.

▶ Click Next.

Figure 9-17
Adding a duration to a date: Step 3

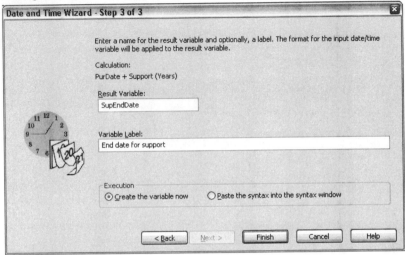

▶ Enter *SupEndDate* for the name of the result variable. Result variables cannot have the same name as an existing variable.

▶ Enter *End date for support* as the label for the result variable. Variable labels for result variables are optional.

▶ Click Finish to create the new variable.

The new variable is displayed in the Data Editor.

Figure 9-18
New variable displayed in Data Editor

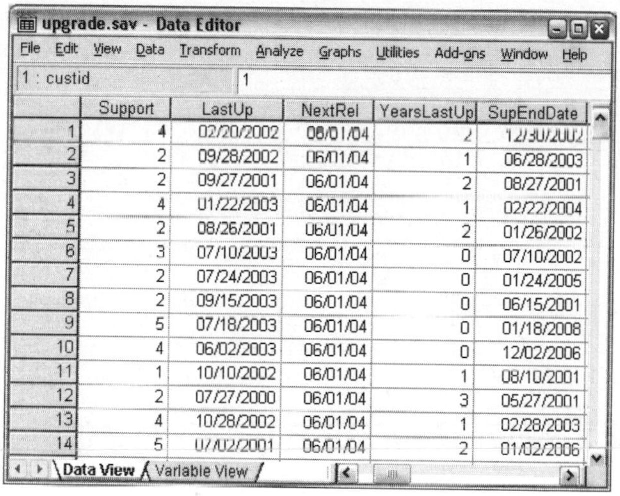

	Support	LastUp	NextRel	YearsLastUp	SupEndDate
1	4	02/20/2002	06/01/04	2	12/30/2002
2	2	09/28/2002	06/01/04	1	06/28/2003
3	2	09/27/2001	06/01/04	2	08/27/2001
4	4	01/22/2003	06/01/04	1	02/22/2004
5	2	08/26/2001	06/01/04	2	01/26/2002
6	3	07/10/2003	06/01/04	0	07/10/2002
7	2	07/24/2003	06/01/04	0	01/24/2005
8	2	09/15/2003	06/01/04	0	06/15/2001
9	5	07/18/2003	06/01/04	0	01/18/2008
10	4	06/02/2003	06/01/04	0	12/02/2006
11	1	10/10/2002	06/01/04	1	08/10/2001
12	2	07/27/2000	06/01/04	3	05/27/2001
13	4	10/28/2002	06/01/04	1	02/28/2003
14	5	07/02/2001	06/01/04	2	01/02/2006

10

Sorting and Selecting Data

Data files are not always organized in the ideal form for your specific needs. To prepare data for analysis, you can select from a wide range of file transformations, including the ability to:

- **Sort data.** You can sort cases based on the value of one or more variables.
- **Select subsets of cases.** You can restrict your analysis to a subset of cases or perform simultaneous analyses on different subsets.

The examples in this chapter use the data file *demo.sav*. For more information, see "Sample Files" in Appendix A on p. 201.

Sorting Data

Sorting cases (sorting rows of the data file) is often useful and sometimes necessary for certain types of analysis.

To reorder the sequence of cases in the data file based on the value of one or more sorting variables:

▶ From the menus choose:
Data
  Sort Cases...

The Sort Cases dialog box is displayed.

Figure 10-1
Sort Cases dialog box

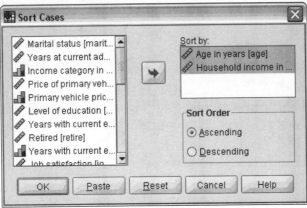

▶ Add the *Age in years [age]* and *Household income in thousands [income]* variables to the Sort by list.

If you select multiple sort variables, the order in which they appear on the Sort by list determines the order in which cases are sorted. In this example, based on the entries in the Sort by list, cases will be sorted by the value of *Household income in thousands [income]* within categories of *Age in years [age]*. For string variables, uppercase letters precede their lowercase counterparts in sort order (for example, the string value *Yes* comes before *yes* in the sort order).

Split-File Processing

To split your data file into separate groups for analysis:

▶ From the menus choose:
Data
  Split File...

The Split File dialog box is displayed.

Figure 10-2
Split File dialog box

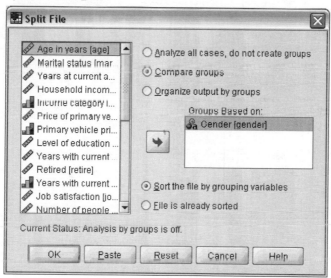

▶ Select **Compare groups** or **Organize output by groups**. (The examples following these steps show the differences between these two options.)

▶ Select *Gender [gender]* to split the file into separate groups for these variables.

You can use numeric, short string, and long string variables as grouping variables. A separate analysis is performed for each subgroup that is defined by the grouping variables. If you select multiple grouping variables, the order in which they appear on the Groups Based on list determines the manner in which cases are grouped.

If you select Compare groups, results from all split-file groups will be included in the same table(s), as shown in the following table of summary statistics that is generated by the Frequencies procedure.

Figure 10-3
Split-file output with single pivot table

Statistics

Household income in thousands

Female	N	Valid	3179
		Missing	0
	Mean		68.7798
	Median		44.0000
	Std. Deviation		75.73510
Male	N	Valid	3221
		Missing	0
	Mean		70.1608
	Median		45.0000
	Std. Deviation		81.56216

If you select Organize output by groups and run the Frequencies procedure, two pivot tables are created: one table for females and one table for males.

Figure 10-4
Split-file output with pivot table for females

Statistics[a]

Household income in thousands

N	Valid	3179
	Missing	0
Mean		68.7798
Median		44.0000
Std. Deviation		75.73510

a. Gender = Female

Figure 10-5
Split-file output with pivot table for males

Statistics[a]

Household income in thousands

N	Valid	3221
	Missing	0
Mean		70.1608
Median		45.0000
Std. Deviation		81.56216

a. Gender = Male

Sorting Cases for Split-File Processing

The Split File procedure creates a new subgroup each time it encounters a different value for one of the grouping variables. Therefore, it is important to sort cases based on the values of the grouping variables before invoking split-file processing.

By default, Split File automatically sorts the data file based on the values of the grouping variables. If the file is already sorted in the proper order, you can save processing time if you select File is already sorted.

Turning Split-File Processing On and Off

After you invoke split-file processing, it remains in effect for the rest of the session unless you turn it off.

- **Analyze all cases.** This option turns split-file processing off.

- **Compare groups** and **Organize output by groups.** This option turns split-file processing on.

If split-file processing is in effect, the message Split File on appears on the status bar at the bottom of the application window.

Selecting Subsets of Cases

You can restrict your analysis to a specific subgroup based on criteria that include variables and complex expressions. You can also select a random sample of cases. The criteria used to define a subgroup can include:

- Variable values and ranges
- Date and time ranges
- Case (row) numbers
- Arithmetic expressions
- Logical expressions
- Functions

To select a subset of cases for analysis:

▶ From the menus choose:
Data
  Select Cases...

This opens the Select Cases dialog box.

Figure 10-6
Select Cases dialog box

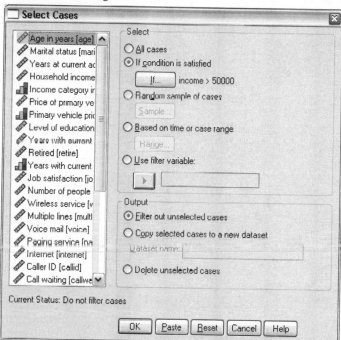

Selecting Cases Based on Conditional Expressions

To select cases based on a conditional expression:

▶ Select If condition is satisfied and click If in the Select Cases dialog box.

This opens the Select Cases If dialog box.

Figure 10-7
Select Cases If dialog box

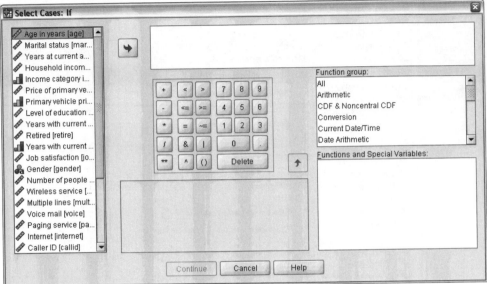

The conditional expression can use existing variable names, constants, arithmetic operators, logical operators, relational operators, and functions. You can type and edit the expression in the text box just like text in an output window. You can also use the calculator pad, variable list, and function list to paste elements into the expression. For more information, see "Using Conditional Expressions" in Chapter 9 on p. 161.

Selecting a Random Sample

To obtain a random sample:

► Select Random sample of cases in the Select Cases dialog box.

► Click Sample.

This opens the Select Cases Random Sample dialog box.

Figure 10-8
Select Cases Random Sample dialog box

You can select one of the following alternatives for sample size:

■ **Approximately.** A user-specified percentage. This option generates a random sample of approximately the specified percentage of cases.

■ **Exactly.** A user-specified number of cases. You must also specify the number of cases from which to generate the sample. This second number should be less than or equal to the total number of cases in the data file. If the number exceeds the total number of cases in the data file, the sample will contain proportionally fewer cases than the requested number.

Selecting a Time Range or Case Range

To select a range of cases based on dates, times, or observation (row) numbers:

▶ Select Based on time or case range and click Range in the Select Cases dialog box.

This opens the Select Cases Range dialog box, in which you can select a range of observation (row) numbers.

Figure 10-9
Select Cases Range dialog box

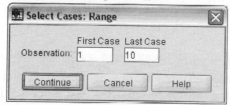

- **First Case.** Enter the starting date and/or time values for the range. If no date variables are defined, enter the starting observation number (row number in the Data Editor, unless Split File is on). If you do not specify a Last Case value, all cases from the starting date/time to the end of the time series are selected.

- **Last Case.** Enter the ending date and/or time values for the range. If no date variables are defined, enter the ending observation number (row number in the Data Editor, unless Split File is on). If you do not specify a First Case value, all cases from the beginning of the time series up to the ending date/time are selected.

For time series data with defined date variables, you can select a range of dates and/or times based on the defined date variables. Each case represents observations at a different time, and the file is sorted in chronological order.

Figure 10-10
Select Cases Range dialog box (time series)

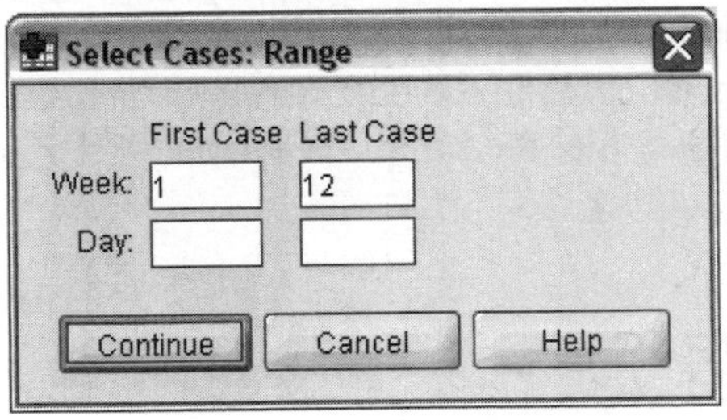

To generate date variables for time series data:

▶ From the menus choose:
Data
  Define Dates...

Treatment of Unselected Cases

You can choose one of the following alternatives for the treatment of unselected cases:

- **Filter out unselected cases.** Unselected cases are not included in the analysis but remain in the dataset. You can use the unselected cases later in the session if you turn filtering off. If you select a random sample or if you select cases based on a conditional expression, this generates a variable named *filter_$* with a value of 1 for selected cases and a value of 0 for unselected cases.

- **Copy selected cases to a new dataset.** Selected cases are copied to a new dataset, leaving the original dataset unaffected. Unselected cases are not included in the new dataset and are left in their original state in the original dataset.

- **Delete unselected cases.** Unselected cases are deleted from the dataset. Deleted cases can be recovered only by exiting from the file without saving any changes and then reopening the file. The deletion of cases is permanent if you save the changes to the data file.

Note: If you delete unselected cases and save the file, the cases cannot be recovered.

Case Selection Status

If you have selected a subset of cases but have not discarded unselected cases, unselected cases are marked in the Data Editor with a diagonal line through the row number.

Figure 10-11
Case selection status

11

Additional Statistical Procedures

This chapter contains brief examples for selected statistical procedures. The procedures are grouped according to the order in which they appear on the Analyze menu.

The examples are designed to illustrate sample specifications that are required to run a statistical procedure. The examples in this chapter use the data file *demo.sav*, with the following exceptions:

- The paired-samples *t* test example uses the data file *dietstudy.sav*, which is a hypothetical data file containing the results of a study of the "Stillman diet." In the examples in this chapter, you must run the procedures to see the output.

- The correlation examples use *Employee data.sav*, which contains historical data about a company's employees.

- The exponential smoothing example uses the data file *inventor.sav*, which contains inventory data that were collected over a period of 70 days.

For information about individual items in a dialog box, click Help. If you want to locate a specific statistic, such as percentiles, use the Index or Search facility in the Help system. For additional information about interpreting the results of these procedures, consult a statistics or data analysis textbook.

Summarizing Data

The Descriptive Statistics submenu on the Analyze menu provides techniques for summarizing data with statistics and charts.

Explore

Suppose that you want to look at the distribution of the years with current employer for each income category. With the Explore procedure, you can examine the distribution of the years with current employer within categories of another variable.

▶ From the menus choose:

Analyze
  Descriptive Statistics
    Explore...

This opens the Explore dialog box.

Figure 11-1
Explore dialog box

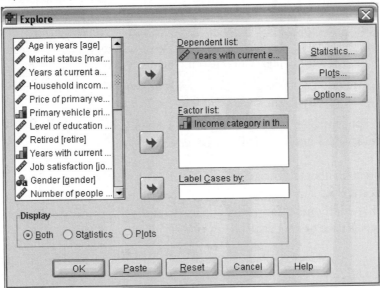

▶ Select *Years with current employer [employ]* and move it to the Dependent List.

▶ Select *Income category in thousands [inccat]* and move it to the Factor List.

▶ Click OK to run the Explore procedure.

In the output, descriptive statistics and a stem-and-leaf plot are displayed for the years with current employer in each income category. The Viewer also contains a boxplot (in standard graphics format) comparing the years with current employer in the income categories. For each category, the boxplot shows the median, interquartile range (25th to 75th percentile), outliers (indicated by O), and extreme values (indicated by *).

More about Summarizing Data

There are many ways to summarize data. For example, to calculate medians or percentiles, use the Frequencies procedure or the Explore procedure. Here are some additional methods:

- **Descriptives.** For income, you can calculate standard scores, sometimes called z scores. Use the Descriptives procedure and select Save standardized values as variables.

- **Crosstabs.** You can use the Crosstabs procedure to display the relationship between two or more categorical variables.

- **Summarize procedure.** You can use the Summarize procedure to write to your output window a listing of the actual values of age, gender, and income of the first 25 or 50 cases.

▶ To run the Summarize procedure, from the menus choose:
Analyze
  Reports
    Case Summaries...

Comparing Means

The Compare Means submenu on the Analyze menu provides techniques for displaying descriptive statistics and testing whether differences are significant between two means for both independent and paired samples. You can also use the One-Way ANOVA procedure to test whether differences are significant among more than two independent means.

Means

In the *demo.sav* file, several variables are available for dividing people into groups. You can then calculate various statistics in order to compare the groups. For example, you can compute the average (mean) household income for males and females. To calculate the means, use the following steps:

▶ From the menus choose:

Analyze
  Compare Means
    Means...

This opens the Means dialog box.

Figure 11-2
Means dialog box (layer 1)

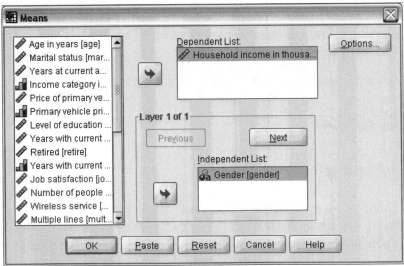

▶ Select *Household income in thousands [income]* and move it to the Dependent List.

▶ Select *Gender [gender]* and move it to the Independent List in layer 1.

▶ Click Next to create another layer.

Figure 11-3
Means dialog box (layer 2)

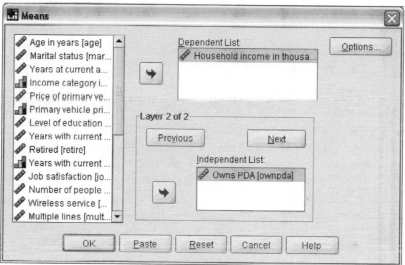

▶ Select *Owns PDA [ownpda]* and move it to the Independent List in layer 2.

▶ Click OK to run the procedure.

Paired-Samples T Test

When the data are structured in such a way that there are two observations on the same individual or observations that are matched by another variable on two individuals (twins, for example), the samples are paired. In the data file *dietstudy.sav*, the beginning and final weights are provided for each person who participated in the study. If the diet worked, we expect that the participant's weight before and after the study would be significantly different.

To carry out a *t* test of the beginning and final weights, use the following steps:

▶ Open the data file *dietstudy.sav*. For more information, see "Sample Files" in Appendix A on p. 201.

▶ From the menus choose:

Analyze
 Compare Means
  Paired-Samples T Test...

This opens the Paired-Samples T Test dialog box.

Figure 11-4
Paired-Samples T Test dialog box

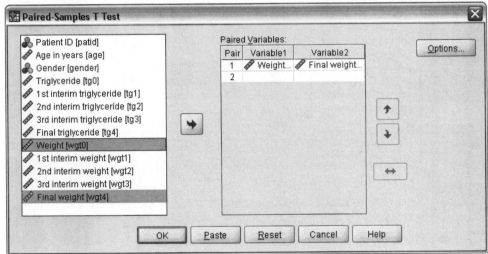

▶ Select *Weight* and *Final weight* as the paired variables.

▶ Click OK to run the procedure.

The results show that the final weight is significantly different from the beginning weight, as indicated by the small probability that is displayed in the *Sig. (2-tailed)* column of the Paired-Samples Test table.

More about Comparing Means

The following examples suggest some ways in which you can use other procedures to compare means.

■ **Independent-Samples T Test.** When you use a *t* test to compare means of one variable across independent groups, the samples are independent. Males and females in the *demo.sav* file can be divided into independent groups by the variable

Gender [gender]. You can use a *t* test to determine whether the mean household incomes of males and females are the same.

- **One-Sample T Test.** You can test whether the household income of people with college degrees differs from a national or state average. Use Select Cases on the Data menu to select the cases with *Level of education [ed]* >= 4. Then, run the One-Sample T Test procedure to compare *Household income in thousands [income]* and the test value 75.

- **One-Way ANOVA.** The variable *Level of education [ed]* divides employees into five independent groups by level of education. You can use the One-Way ANOVA procedure to test whether *Household income in thousands [income]* means for the five groups are significantly different.

ANOVA Models

The General Linear Model submenu on the Analyze menu provides techniques for testing univariate analysis-of-variance models. (If you have only one factor, you can use the One-Way ANOVA procedure on the Compare Means submenu.)

Univariate Analysis of Variance

The GLM Univariate procedure can perform an analysis of variance for factorial designs. A simple factorial design can be used to test whether a person's household income and job satisfaction affect the number of years with the current employer.

▶ From the menus choose:

Analyze
  General Linear Model
    Univariate...

This opens the Univariate dialog box.

Figure 11-5
Univariate dialog box

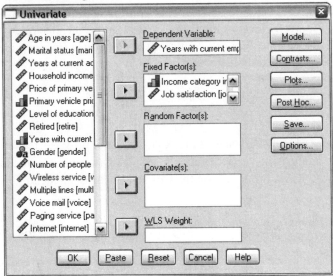

▶ Select *Years with current employer [employ]* and move it to the Dependent Variable list.

▶ Select *Income category in thousands [inccat]* and *Job satisfaction [jobsat]*, and move them to the Fixed Factor(s) list.

▶ Click OK to run the procedure.

In the Tests of Between-Subjects Effects table, you can see that the effects of income and job satisfaction are definitely significant and that the observed significance level of the interaction of income and job satisfaction is 0.000. For further interpretation, consult a statistics or data analysis textbook.

Correlating Variables

The Correlate submenu on the Analyze menu provides measures of association for two or more numeric variables. The examples in this topic use the data file *Employee data.sav*.

Bivariate Correlations

The Bivariate Correlations procedure computes statistics such as Pearson's correlation coefficient. Correlations measure how variables or rank orders are related. Correlation coefficients range in value from –1 (a perfect negative relationship) and +1 (a perfect positive relationship). A value of 0 indicates no linear relationship.

For example, you can use Pearson's correlation coefficient to see if there is a strong linear association between *Current Salary [salary]* and *Beginning Salary [salbegin]* in the data file *Employee data.sav*.

Partial Correlations

The Partial Correlations procedure calculates partial correlation coefficients that describe the relationship between two variables while adjusting for the effects of one or more additional variables.

You can estimate the correlation between *Current Salary [salary]* and *Beginning Salary [salbegin]*, controlling for the linear effects of *Months since Hire [jobtime]* and *Previous Experience [prevexp]*. The number of control variables determines the order of the partial correlation coefficient.

To carry out this Partial Correlations procedure, use the following steps:

▶ Open the *Employee data.sav* file, which is usually in the installation directory.

▶ From the menus choose:

Analyze
  Correlate
    Partial...

This opens the Partial Correlations dialog box.

Figure 11-6
Partial Correlations dialog box

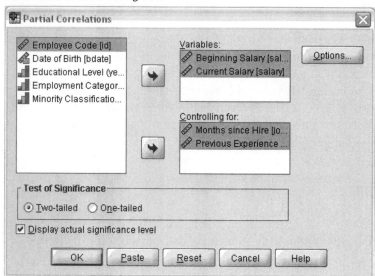

▶ Select *Current Salary [salary]* and *Beginning Salary [salbegin]* and move them to the Variables list.

▶ Select *Months since Hire [jobtime]* and *Previous Experience [prevexp]* and move them to the Controlling for list.

▶ Click OK to run the procedure.

The output shows a table of partial correlation coefficients, the degrees of freedom, and the significance level for the pair *Current Salary [salary]* and *Beginning Salary [salbegin]*.

Regression Analysis

The Regression submenu on the Analyze menu provides regression techniques.

Linear Regression

The Linear Regression procedure examines the relationship between a dependent variable and a set of independent variables. You can use the procedure to predict a person's household income (the dependent variable) based on independent variables such as age, number in household, and years with employer.

▶ From the menus choose:

Analyze
  Regression
    Linear...

This opens the Linear Regression dialog box.

Figure 11-7
Linear Regression dialog box

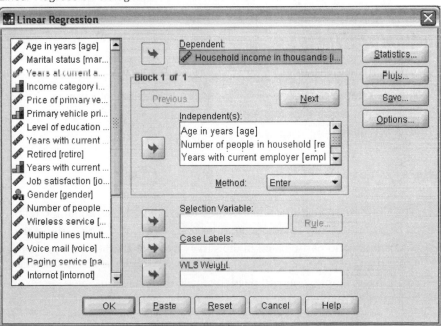

▶ Select *Household income in thousands [income]* and move it to the Dependent list.

▶ Select *Age in years [age]*, *Number of people in household [reside]*, and *Years with current employer [employ]*, and then move them to the Independent(s) list.

▶ Click OK to run the procedure.

The output contains goodness-of-fit statistics and the partial regression coefficients for the variables.

Examining Fit. To see how well the regression model fits your data, you can examine the residuals and other types of diagnostics that this procedure provides. In the Linear Regression dialog box, click Save to see a list of the new variables that you can add to your data file. If you generate any of these variables, they will not be available in a later session unless you save the data file.

Methods. If you have collected a large number of independent variables and want to build a regression model that includes only variables that are statistically related to the dependent variable, you can choose a method from the drop-down list. For example, if you select Stepwise in the above example, only variables that meet the criteria in the Linear Regression Options dialog box are entered in the equation.

Nonparametric Tests

The Nonparametric Tests submenu on the Analyze menu provides nonparametric tests for one sample or for two or more paired or independent samples. Nonparametric tests do not require assumptions about the shape of the distributions from which the data originate.

Chi-Square

The Chi-Square Test procedure is used to test hypotheses about the relative proportion of cases falling into several mutually exclusive groups. You can test the hypothesis that people who participated in the survey occur in the same proportions of gender as the general population (50% males, 50% females).

In this example, you will need to recode the string variable *Gender [gender]* into a numeric variable before you can run the procedure.

▶ From the menus choose:

Transform
  Automatic Recode...

This opens the Automatic Recode dialog box.

Figure 11-8
Automatic Recode dialog box

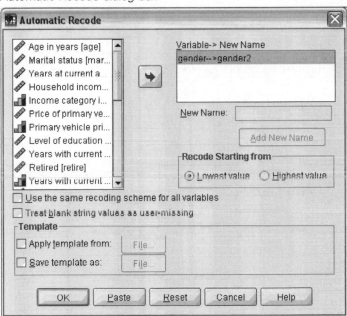

▶ Select the variable *Gender [gender]* and move it to the Variable -> New Name list.

▶ Type gender2 in the New Name text box, and then click the Add New Name button.

▶ Click OK to run the procedure.

This process creates a new numeric variable called *gender2*, which has a value of 1 for females and a value of 2 for males. Now a chi-square test can be run with a numeric variable.

▶ From the menus choose:

Analyze
  Nonparametric Tests
    Chi-Square...

This opens the Chi-Square Test dialog box.

Figure 11-9
Chi-Square Test dialog box

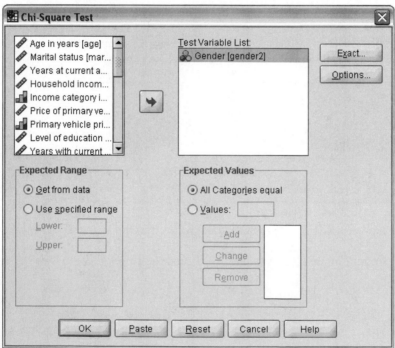

▶ Select *Gender [gender2]* and move it to the Test Variable List.

▶ Select All categories equal because in the general population of working age, the number of males and females is approximately equal.

▶ Click OK to run the procedure.

The output shows a table of the expected and residual values for the categories. The significance of the chi-square test is 0.6. For more information about interpretation of the statistics, consult a statistics or data analysis textbook.

Sample Files

The sample files installed with the product can be found in the *Samples* subdirectory of the installation directory.

Descriptions

Following are brief descriptions of the sample files used in various examples throughout the documentation:

- **accidents.sav.** This is a hypothetical data file that concerns an insurance company that is studying age and gender risk factors for automobile accidents in a given region. Each case corresponds to a cross-classification of age category and gender.

- **adl.sav.** This is a hypothetical data file that concerns efforts to determine the benefits of a proposed type of therapy for stroke patients. Physicians randomly assigned female stroke patients to one of two groups. The first received the standard physical therapy, and the second received an additional emotional therapy. Three months following the treatments, each patient's abilities to perform common activities of daily life were scored as ordinal variables.

- **advert.sav.** This is a hypothetical data file that concerns a retailer's efforts to examine the relationship between money spent on advertising and the resulting sales. To this end, they have collected past sales figures and the associated advertising costs..

- **aflatoxin.sav.** This is a hypothetical data file that concerns the testing of corn crops for aflatoxin, a poison whose concentration varies widely between and within crop yields. A grain processor has received 16 samples from each of 8 crop yields and measured the alfatoxin levels in parts per billion (PPB).

- **aflatoxin20.sav.** This data file contains the aflatoxin measurements from each of the 16 samples from yields 4 and 8 from the *aflatoxin.sav* data file.

- **anorectic.sav.** While working toward a standardized symptomatology of anorectic/bulimic behavior, researchers made a study of 55 adolescents with known eating disorders. Each patient was seen four times over four years, for a total of 220 observations. At each observation, the patients were scored for each of 16 symptoms. Symptom scores are missing for patient 71 at time 2, patient 76 at time 2, and patient 47 at time 3, leaving 217 valid observations.

- **autoaccidents.sav.** This is a hypothetical data file that concerns the efforts of an insurance analyst to model the number of automobile accidents per driver while also accounting for driver age and gender. Each case represents a separate driver and records the driver's gender, age in years, and number of automobile accidents in the last five years.

- **band.sav.** This data file contains hypothetical weekly sales figures of music CDs for a band. Data for three possible predictor variables are also included.

- **bankloan.sav.** This is a hypothetical data file that concerns a bank's efforts to reduce the rate of loan defaults. The file contains financial and demographic information on 850 past and prospective customers. The first 700 cases are customers who were previously given loans. The last 150 cases are prospective customers that the bank needs to classify as good or bad credit risks.

- **bankloan_binning.sav.** This is a hypothetical data file containing financial and demographic information on 5,000 past customers.

- **behavior.sav.** In a classic example , 52 students were asked to rate the combinations of 15 situations and 15 behaviors on a 10-point scale ranging from 0="extremely appropriate" to 9="extremely inappropriate." Averaged over individuals, the values are taken as dissimilarities.

- **behavior_ini.sav.** This data file contains an initial configuration for a two-dimensional solution for *behavior.sav*.

- **brakes.sav.** This is a hypothetical data file that concerns quality control at a factory that produces disc brakes for high-performance automobiles. The data file contains diameter measurements of 16 discs from each of 8 production machines. The target diameter for the brakes is 322 millimeters.

- **breakfast.sav.** In a classic study , 21 Wharton School MBA students and their spouses were asked to rank 15 breakfast items in order of preference with 1="most preferred" to 15="least preferred." Their preferences were recorded under six different scenarios, from "Overall preference" to "Snack, with beverage only."

- **breakfast-overall.sav.** This data file contains the breakfast item preferences for the first scenario, "Overall preference," only.

- **broadband_1.sav.** This is a hypothetical data file containing the number of subscribers, by region, to a national broadband service. The data file contains monthly subscriber numbers for 85 regions over a four-year period.

- **broadband_2.sav.** This data file is identical to *broadband_1.sav* but contains data for three additional months.

- **car_insurance_claims.sav.** A dataset presented and analyzed elsewhere concerns damage claims for cars. The average claim amount can be modeled as having a gamma distribution, using an inverse link function to relate the mean of the dependent variable to a linear combination of the policyholder age, vehicle type, and vehicle age. The number of claims filed can be used as a scaling weight.

- **car_sales.sav.** This data file contains hypothetical sales estimates, list prices, and physical specifications for various makes and models of vehicles. The list prices and physical specifications were obtained alternately from *edmunds.com* and manufacturer sites.

- **carpet.sav.** In a popular example , a company interested in marketing a new carpet cleaner wants to examine the influence of five factors on consumer preference—package design, brand name, price, a *Good Housekeeping* seal, and a money-back guarantee. There are three factor levels for package design, each one differing in the location of the applicator brush; three brand names (*K2R*, *Glory*, and *Bissell*); three price levels; and two levels (either no or yes) for each of the last two factors. Ten consumers rank 22 profiles defined by these factors. The variable *Preference* contains the rank of the average rankings for each profile. Low rankings correspond to high preference. This variable reflects an overall measure of preference for each profile.

- **carpet_prefs.sav.** This data file is based on the same example as described for *carpet.sav*, but it contains the actual rankings collected from each of the 10 consumers. The consumers were asked to rank the 22 product profiles from the most to the least preferred. The variables *PREF1* through *PREF22* contain the identifiers of the associated profiles, as defined in *carpet_plan.sav*.

- **catalog.sav.** This data file contains hypothetical monthly sales figures for three products sold by a catalog company. Data for five possible predictor variables are also included.

- **catalog_seasfac.sav.** This data file is the same as *catalog.sav* except for the addition of a set of seasonal factors calculated from the Seasonal Decomposition procedure along with the accompanying date variables.

- **cellular.sav.** This is a hypothetical data file that concerns a cellular phone company's efforts to reduce churn. Churn propensity scores are applied to accounts, ranging from 0 to 100. Accounts scoring 50 or above may be looking to change providers.

- **ceramics.sav.** This is a hypothetical data file that concerns a manufacturer's efforts to determine whether a new premium alloy has a greater heat resistance than a standard alloy. Each case represents a separate test of one of the alloys; the heat at which the bearing failed is recorded.

- **cereal.sav.** This is a hypothetical data file that concerns a poll of 880 people about their breakfast preferences, also noting their age, gender, marital status, and whether or not they have an active lifestyle (based on whether they exercise at least twice a week). Each case represents a separate respondent.

- **clothing_defects.sav.** This is a hypothetical data file that concerns the quality control process at a clothing factory. From each lot produced at the factory, the inspectors take a sample of clothes and count the number of clothes that are unacceptable.

- **coffee.sav.** This data file pertains to perceived images of six iced-coffee brands . For each of 23 iced-coffee image attributes, people selected all brands that were described by the attribute. The six brands are denoted AA, BB, CC, DD, EE, and FF to preserve confidentiality.

- **contacts.sav.** This is a hypothetical data file that concerns the contact lists for a group of corporate computer sales representatives. Each contact is categorized by the department of the company in which they work and their company ranks. Also recorded are the amount of the last sale made, the time since the last sale, and the size of the contact's company.

- **creditpromo.sav.** This is a hypothetical data file that concerns a department store's efforts to evaluate the effectiveness of a recent credit card promotion. To this end, 500 cardholders were randomly selected. Half received an ad promoting a reduced interest rate on purchases made over the next three months. Half received a standard seasonal ad.

- **customer_dbase.sav.** This is a hypothetical data file that concerns a company's efforts to use the information in its data warehouse to make special offers to customers who are most likely to reply. A subset of the customer base was selected at random and given the special offers, and their responses were recorded.

- **customers_model.sav.** This file contains hypothetical data on individuals targeted by a marketing campaign. These data include demographic information, a summary of purchasing history, and whether or not each individual responded to the campaign. Each case represents a separate individual.

- **customers_new.sav.** This file contains hypothetical data on individuals who are potential candidates for a marketing campaign. These data include demographic information and a summary of purchasing history for each individual. Each case represents a separate individual.

- **debate.sav.** This is a hypothetical data file that concerns paired responses to a survey from attendees of a political debate before and after the debate. Each case corresponds to a separate respondent.

- **debate_aggregate.sav.** This is a hypothetical data file that aggregates the responses in *debate.sav*. Each case corresponds to a cross-classification of preference before and after the debate.

- **demo.sav.** This is a hypothetical data file that concerns a purchased customer database, for the purpose of mailing monthly offers. Whether or not the customer responded to the offer is recorded, along with various demographic information.

- **demo_cs_1.sav.** This is a hypothetical data file that concerns the first step of a company's efforts to compile a database of survey information. Each case corresponds to a different city, and the region, province, district, and city identification are recorded.

- **demo_cs_2.sav.** This is a hypothetical data file that concerns the second step of a company's efforts to compile a database of survey information. Each case corresponds to a different household unit from cities selected in the first step, and the region, province, district, city, subdivision, and unit identification are recorded. The sampling information from the first two stages of the design is also included.

- **demo_cs.sav.** This is a hypothetical data file that contains survey information collected using a complex sampling design. Each case corresponds to a different household unit, and various demographic and sampling information is recorded.

- **dietstudy.sav.** This hypothetical data file contains the results of a study of the "Stillman diet" . Each case corresponds to a separate subject and records his or her pre- and post-diet weights in pounds and triglyceride levels in mg/100 ml.

- **dischargedata.sav.** This is a data file concerning *Seasonal Patterns of Winnipeg Hospital Use*, from the Manitoba Centre for Health Policy.

- **dvdplayer.sav.** This is a hypothetical data file that concerns the development of a new DVD player. Using a prototype, the marketing team has collected focus group data. Each case corresponds to a separate surveyed user and records some demographic information about them and their responses to questions about the prototype.

- **flying.sav.** This data file contains the flying mileages between 10 American cities.

- **german_credit.sav.** This data file is taken from the "German credit" dataset in the Repository of Machine Learning Databases at the University of California, Irvine.

- **grocery_1month.sav.** This hypothetical data file is the *grocery_coupons.sav* data file with the weekly purchases "rolled-up" so that each case corresponds to a separate customer. Some of the variables that changed weekly disappear as a result, and the amount spent recorded is now the sum of the amounts spent during the four weeks of the study.

- **grocery_coupons.sav.** This is a hypothetical data file that contains survey data collected by a grocery store chain interested in the purchasing habits of their customers. Each customer is followed for four weeks, and each case corresponds to a separate customer-week and records information about where and how the customer shops, including how much was spent on groceries during that week.

- **guttman.sav.** Bell presented a table to illustrate possible social groups. Guttman used a portion of this table, in which five variables describing such things as social interaction, feelings of belonging to a group, physical proximity of members, and formality of the relationship were crossed with seven theoretical social groups, including crowds (for example, people at a football game), audiences (for example, people at a theater or classroom lecture), public (for example, newspaper or television audiences), mobs (like a crowd but with much more intense interaction), primary groups (intimate), secondary groups (voluntary), and the modern community (loose confederation resulting from close physical proximity and a need for specialized services).

- **healthplans.sav.** This is a hypothetical data file that concerns an insurance group's efforts to evaluate four different health care plans for small employers. Twelve employers are recruited to rank the plans by how much they would prefer to offer them to their employees. Each case corresponds to a separate employer and records the reactions to each plan.

- **health_funding.sav.** This is a hypothetical data file that contains data on health care funding (amount per 100 population), disease rates (rate per 10,000 population), and visits to health care providers (rate per 10,000 population). Each case represents a different city.

- **hivassay.sav.** This is a hypothetical data file that concerns the efforts of a pharmaceutical lab to develop a rapid assay for detecting HIV infection. The results of the assay are eight deepening shades of red, with deeper shades indicating greater likelihood of infection. A laboratory trial was conducted on 2,000 blood samples, half of which were infected with HIV and half of which were clean.

- **hourlywagedata.sav.** This is a hypothetical data file that concerns the hourly wages of nurses from office and hospital positions and with varying levels of experience.

- **insure.sav.** This is a hypothetical data file that concerns an insurance company that is studying the risk factors that indicate whether a client will have to make a claim on a 10-year term life insurance contract. Each case in the data file represents a pair of contracts, one of which recorded a claim and the other didn't, matched on age and gender.

- **judges.sav.** This is a hypothetical data file that concerns the scores given by trained judges (plus one enthusiast) to 300 gymnastics performances. Each row represents a separate performance; the judges viewed the same performances.

- **kinship_dat.sav.** Rosenberg and Kim set out to analyze 15 kinship terms (aunt, brother, cousin, daughter, father, granddaughter, grandfather, grandmother, grandson, mother, nephew, niece, sister, son, uncle). They asked four groups of college students (two female, two male) to sort these terms on the basis of similarities. Two groups (one female, one male) were asked to sort twice, with the second sorting based on a different criterion from the first sort. Thus, a total of six "sources" were obtained. Each source corresponds to a 15 × 15 proximity matrix, whose cells are equal to the number of people in a source minus the number of times the objects were partitioned together in that source.

- **kinship ini.sav.** This data file contains an initial configuration for a three-dimensional solution for *kinship_dat.sav*.

- **kinship_var.sav.** This data file contains independent variables *gender*, *gener*(ation), and *degree* (of separation) that can be used to interpret the dimensions of a solution for *kinship_dat.sav*. Specifically, they can be used to restrict the space of the solution to a linear combination of these variables.

- **mailresponse.sav.** This is a hypothetical data file that concerns the efforts of a clothing manufacturer to determine whether using first class postage for direct mailings results in faster responses than bulk mail. Order-takers record how many weeks after the mailing each order is taken.

- **marketvalues.sav.** This data file concerns home sales in a new housing development in Algonquin, Ill., during the years from 1999–2000. These sales are a matter of public record.

- **mutualfund.sav.** This data file concerns stock market information for various tech stocks listed on the S&P 500. Each case corresponds to a separate company.

- **nhis2000_subset.sav.** The National Health Interview Survey (NHIS) is a large, population-based survey of the U.S. civilian population. Interviews are carried out face-to-face in a nationally representative sample of households. Demographic information and observations about health behaviors and status are obtained for members of each household. This data file contains a subset of information from the 2000 survey. National Center for Health Statistics. National Health Interview Survey, 2000. Public-use data file and documentation. *ftp://ftp.cdc.gov/pub/Health_Statistics/NCHS/Datasets/NHIS/2000/*. Accessed 2003.

- **ozone.sav.** The data include 330 observations on six meteorological variables for predicting ozone concentration from the remaining variables. Previous researchers , , among others found nonlinearities among these variables, which hinder standard regression approaches.

- **pain_medication.sav.** This hypothetical data file contains the results of a clinical trial for anti-inflammatory medication for treating chronic arthritic pain. Of particular interest is the time it takes for the drug to take effect and how it compares to an existing medication.

- **patient_los.sav.** This hypothetical data file contains the treatment records of patients who were admitted to the hospital for suspected myocardial infarction (MI, or "heart attack"). Each case corresponds to a separate patient and records many variables related to their hospital stay.

- **patlos_sample.sav.** This hypothetical data file contains the treatment records of a sample of patients who received thrombolytics during treatment for myocardial infarction (MI, or "heart attack"). Each case corresponds to a separate patient and records many variables related to their hospital stay.

- **polishing.sav.** This is the "Nambeware Polishing Times" data file from the Data and Story Library. It concerns the efforts of a metal tableware manufacturer (Nambe Mills, Santa Fe, N. M.) to plan its production schedule. Each case represents a different item in the product line. The diameter, polishing time, price, and product type are recorded for each item.

- **poll_cs.sav.** This is a hypothetical data file that concerns pollsters' efforts to determine the level of public support for a bill before the legislature. The cases correspond to registered voters. Each case records the county, township, and neighborhood in which the voter lives.

- **poll_cs_sample.sav.** This hypothetical data file contains a sample of the voters listed in *poll_cs.sav*. The sample was taken according to the design specified in the *poll.csplan* plan file, and this data file records the inclusion probabilities and sample weights. Note, however, that because the sampling plan makes use of a probability-proportional-to-size (PPS) method, there is also a file containing the joint selection probabilities (*poll_jointprob.sav*). The additional variables corresponding to voter demographics and their opinion on the proposed bill were collected and added the data file after the sample as taken.

- **property_assess.sav.** This is a hypothetical data file that concerns a county assessor's efforts to keep property value assessments up to date on limited resources. The cases correspond to properties sold in the county in the past year. Each case in the data file records the township in which the property lies, the assessor who last visited the property, the time since that assessment, the valuation made at that time, and the sale value of the property.

- **property_assess_cs.sav.** This is a hypothetical data file that concerns a state assessor's efforts to keep property value assessments up to date on limited resources. The cases correspond to properties in the state. Each case in the data file records the county, township, and neighborhood in which the property lies, the time since the last assessment, and the valuation made at that time.

- **property_assess_cs_sample.sav.** This hypothetical data file contains a sample of the properties listed in *property_assess_cs.sav*. The sample was taken according to the design specified in the *property_assess.csplan* plan file, and this data file records the inclusion probabilities and sample weights. The additional variable *Current value* was collected and added to the data file after the sample was taken.

- **recidivism.sav.** This is a hypothetical data file that concerns a government law enforcement agency's efforts to understand recidivism rates in their area of jurisdiction. Each case corresponds to a previous offender and records their demographic information, some details of their first crime, and then the time until their second arrest, if it occurred within two years of the first arrest.

- **salesperformance.sav.** This is a hypothetical data file that concerns the evaluation of two new sales training courses. Sixty employees, divided into three groups, all receive standard training. In addition, group 2 gets technical training; group 3, a hands-on tutorial. Each employee was tested at the end of the training course and their score recorded. Each case in the data file represents a separate trainee and records the group to which they were assigned and the score they received on the exam.

- **satisf.sav.** This is a hypothetical data file that concerns a satisfaction survey conducted by a retail company at 4 store locations. 582 customers were surveyed in all, and each case represents the responses from a single customer.

- **screws.sav.** This data file contains information on the characteristics of screws, bolts, nuts, and tacks .

- **shampoo_ph.sav.** This is a hypothetical data file that concerns the quality control at a factory for hair products. At regular time intervals, six separate output batches are measured and their pH recorded. The target range is 4.5–5.5.

- **ships.sav.** A dataset presented and analyzed elsewhere that concerns damage to cargo ships caused by waves. The incident counts can be modeled as occurring at a Poisson rate given the ship type, construction period, and service period. The aggregate months of service for each cell of the table formed by the cross-classification of factors provides values for the exposure to risk.

- **site.sav.** This is a hypothetical data file that concerns a company's efforts to choose new sites for their expanding business. They have hired two consultants to separately evaluate the sites, who, in addition to an extended report, summarized each site as a "good," "fair," or "poor" prospect.

- **siteratings.sav.** This is a hypothetical data file that concerns the beta testing of an e-commerce firm's new Web site. Each case represents a separate beta tester, who scored the usability of the site on a scale from 0–20.

- **smokers.sav.** This data file is abstracted from the 1998 National Household Survey of Drug Abuse and is a probability sample of American households. Thus, the first step in an analysis of this data file should be to weight the data to reflect population trends.

- **smoking.sav.** This is a hypothetical table introduced by Greenacre . The table of interest is formed by the crosstabulation of smoking behavior by job category. The variable *Staff Group* contains the job categories *Sr Managers*, *Jr Managers*, *Sr*

Employees, *Jr Employees*, and *Secretaries*, plus the category *National Average*, which can be used as supplementary to an analysis. The variable *Smoking* contains the behaviors *None*, *Light*, *Medium*, and *Heavy*, plus the categories *No Alcohol* and *Alcohol*, which can be used as supplementary to an analysis.

- **storebrand.sav.** This is a hypothetical data file that concerns a grocery store manager's efforts to increase sales of the store brand detergent relative to other brands. She puts together an in-store promotion and talks with customers at check-out. Each case represents a separate customer.

- **stores.sav.** This data file contains hypothetical monthly market share data for two competing grocery stores. Each case represents the market share data for a given month.

- **stroke_clean.sav.** This hypothetical data file contains the state of a medical database after it has been cleaned using procedures in the Data Preparation option.

- **stroke_invalid.sav.** This hypothetical data file contains the initial state of a medical database and contains several data entry errors.

- **stroke_valid.sav.** This hypothetical data file contains the state of a medical database after the values have been checked using the Validate Data procedure. It still contains potentially anomalous cases.

- **tastetest.sav.** This is a hypothetical data file that concerns the effect of mulch color on the taste of crops. Strawberries grown in red, blue, and black mulch were rated by taste-testers on an ordinal scale of 1 to 5 (far below to far above average). Each case represents a separate taste-tester.

- **telco.sav.** This is a hypothetical data file that concerns a telecommunications company's efforts to reduce churn in their customer base. Each case corresponds to a separate customer and records various demographic and service usage information.

- **telco_extra.sav.** This data file is similar to the *telco.sav* data file, but the "tenure" and log-transformed customer spending variables have been removed and replaced by standardized log-transformed customer spending variables.

- **telco_missing.sav.** This data file is the same as the *telco_mva_complete.sav* data file, but some of the data have been replaced with missing values.

- **telco_mva_complete.sav.** This data file is a subset of the *telco.sav* data file but with different variable names.

- **testmarket.sav.** This hypothetical data file concerns a fast food chain's plans to add a new item to its menu. There are three possible campaigns for promoting the new product, so the new item is introduced at locations in several randomly selected markets. A different promotion is used at each location, and the weekly sales of the new item are recorded for the first four weeks. Each case corresponds to a separate location-week.

- **testmarket_1month.sav.** This hypothetical data file is the *testmarket.sav* data file with the weekly sales "rolled-up" so that each case corresponds to a separate location. Some of the variables that changed weekly disappear as a result, and the sales recorded is now the sum of the sales during the four weeks of the study.

- **tree_car.sav.** This is a hypothetical data file containing demographic and vehicle purchase price data.

- **tree_credit.sav.** This is a hypothetical data file containing demographic and bank loan history data.

- **tree_missing_data.sav** This is a hypothetical data file containing demographic and bank loan history data with a large number of missing values.

- **tree_score_car.sav.** This is a hypothetical data file containing demographic and vehicle purchase price data.

- **tree_textdata.sav.** A simple data file with only two variables intended primarily to show the default state of variables prior to assignment of measurement level and value labels.

- **tv-survey.sav.** This is a hypothetical data file that concerns a survey conducted by a TV studio that is considering whether to extend the run of a successful program. 906 respondents were asked whether they would watch the program under various conditions. Each row represents a separate respondent; each column is a separate condition.

- **ulcer_recurrence.sav.** This file contains partial information from a study designed to compare the efficacy of two therapies for preventing the recurrence of ulcers. It provides a good example of interval-censored data and has been presented and analyzed elsewhere .

- **ulcer_recurrence_recoded.sav.** This file reorganizes the information in *ulcer_recurrence.sav* to allow you model the event probability for each interval of the study rather than simply the end-of-study event probability. It has been presented and analyzed elsewhere .

■ **verd1985.sav.** This data file concerns a survey . The responses of 15 subjects to 8 variables were recorded. The variables of interest are divided into three sets. Set 1 includes *age* and *marital*, set 2 includes *pet* and *news*, and set 3 includes *music* and *live*. *Pet* is scaled as multiple nominal and *age* is scaled as ordinal; all of the other variables are scaled as single nominal.

■ **virus.sav.** This is a hypothetical data file that concerns the efforts of an Internet service provider (ISP) to determine the effects of a virus on its networks. They have tracked the (approximate) percentage of infected e-mail traffic on its networks over time, from the moment of discovery until the threat was contained.

■ **waittimes.sav.** This is a hypothetical data file that concerns customer waiting times for service at three different branches of a local bank. Each case corresponds to a separate customer and records the time spent waiting and the branch at which they were conducting their business.

■ **webusability.sav.** This is a hypothetical data file that concerns usability testing of a new e-store. Each case corresponds to one of five usability testers and records whether or not the tester succeeded at each of six separate tasks.

■ **wheeze_steubenville.sav.** This is a subset from a longitudinal study of the health effects of air pollution on children . The data contain repeated binary measures of the wheezing status for children from Steubenville, Ohio, at ages 7, 8, 9 and 10 years, along with a fixed recording of whether or not the mother was a smoker during the first year of the study.

■ **workprog.sav.** This is a hypothetical data file that concerns a government works program that tries to place disadvantaged people into better jobs. A sample of potential program participants were followed, some of whom were randomly selected for enrollment in the program, while others were not. Each case represents a separate program participant.